MAGNIFICENT
MARS

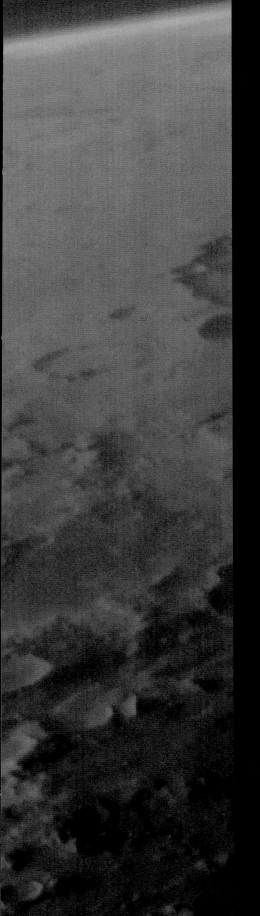

MAGNIF

MA

KEN CRO

FREE PRESS
A Division of Simon & Schuster, Inc.
1230 Avenue of the Americas
New York, NY 10020

*f*P

Designed by Vertigo Design, NYC

Printed and bound in Great Britain by Butler & Tanner, Limited

10   9   8   7   6   5   4   3   2   1

Library of Congress Cataloging-in-Publication Data

Croswell, Ken.
  Magnificent Mars / Ken Croswell.
    p. cm.
  Includes bibliographical references and index.
   ISBN 0-7432-2601-1 (hc)
   1. Mars (Planet) I. Title.

QB641.C76 2003
523.43—dc21   2003049123

ISBN 0-7432-2601-1

For information regarding special discounts for bulk
purchases, please contact Simon & Schuster Special Sales at
1-800-456-6798 or business@simonandschuster.com

CONTENTS

MAGNIFICENT
MARS

Welcome to Mars

Welcome to Mars. *Magnificent Mars* is your ticket, a journey in words and stunning pictures that explore the red planet from pole to pole. With them you climb atop mighty volcanoes that dwarf Mount Everest, snake down giant canyons that could stretch from Ohio to California, dig through thick ice that caps the planet's poles, even poke through craters that puncture the moons of Mars.

This neighbor world may hold the key to whether life abounds throughout the universe. On Earth life arose and flourished, but there's no guarantee it did so elsewhere. In its youth, however, Mars was wetter and probably warmer. If ancient Mars also gave rise to life, then many other worlds in the cosmos have surely done the same.

Furthermore, Mars offers the chance to study a planet's history in a way that Earth doesn't. Terrestrial oceans, rainfall, continental drift, and volcanic eruptions have largely erased the Earth's distant past, whereas much of the Martian surface preserves a record of the ancient era when life was struggling to arise on Earth—and possibly Mars.

Magnificent Mars centers around the four elements of Mars: EARTH, AIR, FIRE, and WATER.

EARTH explores Martian geology, from the planet's iron core to its rocky mantle and surface, unfolding new, rainbow-colored topographic maps that show the striking dichotomy between the planet's smooth northern plains and its cratered southern highlands. Spacecraft have landed on Mars and witnessed the surface close up, while meteorites from the planet allow scientists to analyze Martian rocks in the laboratory.

AIR describes the Martian atmosphere, thin and cold, which nevertheless whips up ferocious dust storms that envelop the globe. The air's present composition offers clues to its past, revealing that long ago the atmosphere was thicker, laden with greenhouse gases such as carbon dioxide and water vapor that warmed the world below.

FIRE explores the red planet's volcanoes, the tallest mountains in the solar system. In their youth, the volcanoes not only flooded much of Mars with lava, but they also emitted the greenhouse gases that warmed the planet. On rare occasions, the volcanoes still erupt: spacecraft images reveal lava flows only a few million years old, and most of the Martian meteorites are young volcanic rocks. When the largest volcanic province arose, it cracked the planet's crust and created Valles Marineris, the largest canyons in the solar system.

thought to be crucial to life. All terrestrial life requires liquid water, and ancient Mars had rivers, likely lakes, and possibly even an ocean. Furthermore, enormous floods carved channels whose remains testify to the water's fury. Perhaps, billions of years ago, a pool of water, stirred by winds and warmed by nearby volcanoes, shuffled its chemicals in just the right way to give birth to the first living beings in the solar system; their fossils may still be preserved in the ruddy Martian soil, waiting for the first Mars-bound astronauts to discover.

NASA spacecraft have returned thousands of images of Mars and its moons. In choosing the very best, I selected images from these spacecraft: Viking 1 Orbiter, Viking 1 Lander, Viking 2 Orbiter, Viking 2 Lander, Mars Pathfinder, Mars Global Surveyor, Mars Odyssey, the Hubble Space Telescope, Apollo 17, Galileo, SOHO, and the Soviet Phobos mission. Expert astrophotographer Tony Hallas then digitally reprocessed these images to bring out the very best color and sharpness, with the ambitious goal of making them look even better than NASA's own. I thank those who helped me obtain the highest resolution images, which are vital to such a large-format book: Antoinette Beiser, Michael Caplinger, Xaviant Ford, Peter Neivert, Greg Neumann, Mark Robinson, Damon Simonelli, Deborah Lee Soltesz, Peter Thomas, Adrienne Wasserman; and NASA, the Jet Propulsion Laboratory, the U.S. Geological Survey, and Lowell Observatory.

Magnificent Mars centers around the four elements of Mars: EARTH, AIR, FIRE and WATER

Numerous scientists provided me with insights about Mars. I thank Mario Acuña, Victor Baker, Michael Carr, Philip Christensen, Kenneth Edgett, Robert Haberle, William K. Hartmann, James Head, Bruce Jakosky, James Kasting, Christopher McKay, Jeffrey Moore, Tobias Owen, Timothy Parker, Robert Pepin, Damon Simonelli, David Smith, and Charles Yoder.

I thank those who read the manuscript in its entirety and offered their comments: Alex Blackwell, David Hudgins, and Richard Pogge.

I thank my acquiring editor, Stephen Morrow, for his enthusiasm for this ambitious project; my current editor, Andrea Au, for her diligence in executing it; my copyediting supervisor, Loretta Denner, for her exactitude; and my senior production manager, Peter McCulloch, for seeing that the images reproduce as beautifully as those in *Magnificent Universe*. Finally, and especially, I thank my agent, Russell Galen, for his support of my work.

Mars Through the Centuries

A BRILLIANT RED BEACON, Mars gleams across an interplanetary gulf measured in millions of miles, distinct from its peers by virtue of its striking color. The hue arises from a rusty, dusty desert draping most of the globe; bright white caps of ice deck the poles. As Mars spins, observers peering through telescopes can watch surface features rotate into and out of view. Centuries before the first Mars-bound spacecraft, these observations painted a picture of an intriguing neighbor world, possibly inhabited, on the Earth's planetary doorstep.

FACING PAGE: Long before spacecraft reached Mars, telescopic observers discovered features that appear in this Hubble Space Telescope image. The first was Syrtis Major, the dark triangular patch just right of center. Because most telescopes invert views, this image is upside down—south is up, north is down. The north polar cap is at bottom, and an apparent polar cap at top is really a giant frosty crater named Hellas.

Mars is the fourth planet from the Sun, next out from Earth and last of the small, rocky worlds that rule the inner solar system; beyond is the asteroid belt that separates the inner worlds from the giant planets. Mars glides through space at a mean distance from the Sun of 142 million miles—52 percent greater than the Earth's distance of 93 million miles. Therefore, the Martian sky features a Sun that looks two thirds as large as it does from Earth, and sunlight strikes Mars with 43 percent of its terrestrial strength.

Because Mars resides beyond the Earth's orbit, its path around the Sun is longer and its speed slower. Mars therefore takes more time to revolve around the Sun: 687 days, or 1.88 Earth years. This is the Martian year. A Martian who is sixty Earth years old would have celebrated only thirty-two birthdays on Mars.

Both the Earth and Mars dance with Venus. As the Earth orbits the Sun five times, Venus does so eight times. Therefore, an Earthling who sees Venus at a particular spot—say, in the west two hours after sunset—will see it in the same spot eight years later. The Mars-Venus dance lacks this precision, but Mars nevertheless completes one orbit around the Sun in nearly the same time that Venus completes three.

Half the diameter of Earth—only 4,221 miles across—Mars sports just a quarter as much surface area. Nevertheless, Martian land area equals the Earth's, because oceans inundate nearly three fourths of the terrestrial globe. The mass of Mars amounts to only 11 percent of the Earth's, so to build one Earth you would need nine Mars-sized chunks. Likewise, Martian gravity is weaker than terrestrial. If you weigh 100 on Earth, you'd weigh only 38 on Mars.

A day on Mars could pass for its terrestrial equivalent, because Mars spins once every 24 hours and 37 minutes. No other planet in the solar system has such an earthly rotation period. Mars spins about an axis tilted like the Earth's—25.2 degrees, versus the Earth's 23.4 degrees. No other planet in the solar system has such an earthly axial tilt. This tilt causes seasons on Mars that resemble those on Earth. When the northern hemisphere tilts toward the Sun, it warms and experiences summer; when the northern hemisphere tilts away, it cools and experiences winter. Because the Martian year lasts about twice as long as the terrestrial, however, Martian seasons also last about twice as long as their terrestrial counterparts. Like the Earth, Mars even has a satellite—actually, two of them, circling the planet just as the Moon circles the Earth. Unlike the Moon, however, each is tiny, just a few miles across.

Because of its terrestrial similarities, Mars has long inspired visions of a living world close to home. In the past, these visions were of life in the present tense; at the present, they are of life in the past tense. Dried-up riverbeds cutting through ancient terrain in southern Mars suggest that the world was once warm and wet. Furthermore, much of the northern hemisphere is smooth and depressed, hinting that it once hosted an ocean. Billions of years ago, then, Mars may have been friendlier to life.

It's certainly not today. The planet's mean temperature is 67 degrees Fahrenheit below zero—as frigid as the mean *winter* temperature at the Earth's south pole—so the liquid water that nourishes life on Earth is absent from Mars. The little air is mostly carbon dioxide, the same gas we get rid of when we breathe. Good luck finding any oxygen to breathe or any ozone to protect you from the Sun's lethal ultraviolet rays.

Nevertheless, aside from the Earth, Mars is the friendliest planet in the solar system. It would be the easiest for a properly suited astronaut to survive on: it offers a solid surface, a thin atmosphere, and a temperature that at times approaches the pleasant. Although Martian air is thin, it is billions of times thicker than the Moon's; although Mars lacks liquid water, it does possess frozen water; and although Mars is cold, summer temperatures flirt with 80 degrees Fahrenheit *above* zero. As a result, Mars will surely be the first extraterrestrial planet to welcome human settlers.

A Heavenly Sight

AT ITS BEST, Mars outshines every celestial body but the Sun, the Moon, Venus, and Jupiter. At its very best, Mars edges out Jupiter, briefly becoming the fourth brightest celestial object of all. At its worst, however, the red planet looks like just another star.

Mars is brightest when it's nearest Earth. The red world swings closer than any other planet but Venus. Venus comes as close as 25 million miles, whereas Mars can pass within 35 million. Close approaches occur approximately every two years and two months, when the planet is on the opposite side of the Earth from the Sun, a configuration called opposition. During opposition a Mars observer looks away from the Sun: the planet rises around sunset, climbs highest at midnight, and sets around sunrise, so astronomers can view it the entire night.

All oppositions are not created equal, however. During the best oppositions, Mars looks nearly twice as large and over four times as bright as during the worst. That's because Mars follows a fairly elliptical path around the Sun. If opposition happens to occur when Mars is farthest from the Sun, then Mars is also far from the Earth, hampering observations. On the other hand, if opposition coincides with Mars's closest approach to the Sun, then the planet is also close to the Earth, the ideal situation for terrestrial observers.

A number called eccentricity shows how elliptical the Martian orbit is. Eccentricity ranges from 0 percent, for a circular orbit, to nearly 100 percent, for the most elliptical orbit possible. The planet boasting the most circular orbit is Venus, with an orbital eccentricity of just 0.7 percent. To the eye Venus's orbit looks like a perfect circle. So does the Earth's, whose eccentricity is only 1.7 percent. Over a year, the Earth's distance from the Sun varies by 3 million miles—from 91.4 million to 94.5 million miles.

Mars will surely be the first extraterrestrial planet to welcome human settlers

In contrast, the orbital eccentricity of Mars is over five times the Earth's, 9.3 percent. Even to the eye the Martian orbit looks lopsided. During a Martian year the planet's distance from the Sun varies by over 26 million miles—from 128 million to 155 million miles. The only planets in the solar system that pursue more elliptical courses are Mercury and Pluto. Perhaps not coincidentally they are also the only planets that are smaller.

FOLLOWING PAGES: Earth and Mars to true relative scale: Mars is about half the diameter of Earth. North is up.

Because Mars has such an elliptical orbit, its opposition distance from Earth can amount to as little as 35 million miles or as much as 63 million miles, nearly twice as far. Although Martian oppositions occur every 2 years, especially favorable oppositions come only once every 15 or 17 years, when Mars skirts closest to Earth and looks its best. These favorable oppositions give observers their best chance to scrutinize Mars and make major discoveries. For example, during one such opposition, astronomers first sighted the two Martian moons.

The last favorable Martian opposition occurred in August 2003; the next will be in July 2018 and September 2035. As these examples illustrate, favorable Martian oppositions occur when the Earth's northern hemisphere is having summer or early autumn—an unfortunate situation for northern observers, since it means Mars is close to the horizon. That's because in summer the zodiac constellations opposite the Sun, which Mars traverses at opposition, are south of the celestial equator and thus harder to observe from the north.

Mars

from Afar

THE ANCIENTS knew only five planets—Mercury, Venus, Mars, Jupiter, Saturn—which stood out from the stars by wandering against the starry backdrop. Even then, Mars further stood out through its vivid hue, "a blazing ruby in the midnight sky," as Owen Gingerich, of the Harvard-Smithsonian Center for Astrophysics, has called it. To ancient people the planet's red color connoted blood. The Egyptians

named it "the Red One," the Babylonians "the Star of Death," and the Greeks "Ares" for their god of war, whose Roman equivalent was Mars. The planet's symbol was ♂, a shield and spear. When Mars was bright, said astrologers, evil prevailed. That one of the brightest and most beautiful planets bears the name of the god of war would tell an extraterrestrial civilization most of what it needed to know about human history.

Mars lent his name elsewhere, to the month of March as well as the word "martial." Mars also ruled one of the days of the week, Tuesday, for Tiu was the Norse god of war.

In more modern times, Mars starred in the revolution of the revolutions. Even those astronomers who had adopted Nicolaus Copernicus's radical Sun-centered solar system believed that the planets revolved on circular orbits. German astronomer—and astrologer—Johannes Kepler realized otherwise. Kepler had inherited observations from the great Danish astronomer Tycho Brahe and for years had tried various schemes to match Tycho's observations of the red planet's position as it traveled through the constellations, but he failed every time. At one point, however, Kepler almost succeeded. He tried a circular orbit on which Mars sped up and slowed down. At its worst, this theory predicted a position for Mars that deviated from Tycho's observations by only 8 arcminutes—a mere 2/15 degree. A lesser scientist would have ignored the discrepancy or even fudged the theory to make it fit. But appreciating the quality of Tycho's observations, Kepler instead threw out his otherwise promising theory.

In 1605, Kepler hit upon the truth: Mars follows an *elliptical* path around the Sun, speeding up when closest and slowing down when farthest. "In order to be able to arrive at understanding," Kepler wrote, "it was absolutely necessary to take the motion of Mars as the basis; otherwise, these secrets would have remained eternally hidden." The other easy-to-observe planets—Venus, Jupiter, Saturn—also obey Kepler's laws, but on paths so circular that Kepler could not have used them to make his great discovery. Mars was no longer the harbinger of doom but instead the revealer of universal laws.

Kepler published his discovery in 1609, the momentous year when the great Italian astronomer Galileo Galilei first turned a small telescope on the heavens. Surprisingly, Galileo's telescope did little for Mars. Although Galileo saw spots on the Sun, craters on the Moon, the phases of Venus, a quartet of moons around Jupiter, and some oddity about Saturn, he saw nothing on Mars. The problem was Mars itself. It's small, at best appearing no larger than a modest lunar crater.

In 1659, Dutch physicist and astronomer Christiaan Huygens produced what Percival Lowell later called "the first drawing of Mars worthy the name ever made by man." Huygens had achieved fame with a more distant planet, spotting Saturn's largest moon and discovering that its oddity was really a ring. Now, on his Martian map, he depicted a dark triangular region, later named Syrtis Major, for a Libyan gulf. By observing Syrtis Major's position from night to night, Huygens was the first to deduce that Mars spun with a period of about twenty-four hours, like the Earth.

Seven years later, in 1666, Italian astronomer Giovanni Cassini discovered that the poles of Mars were white. He refined the Martian rotation period to 24 hours, 40 minutes—just 3 minutes too long. Cassini suggested that Syrtis Major and the other dark areas on Mars were seas and the light areas continents. Half a century later, in 1719, Cassini's nephew, Giacomo Filippo Maraldi, watched the white spots at the poles wax and wane with the passing of the Martian seasons.

Another half century onward, the great German-born English astronomer William Herschel—

Mars was no longer the harbinger of doom but instead the revealer of universal laws

discoverer of the planet Uranus in 1781—speculated that the white spots were polar ice caps, melting and freezing in response to the changing seasons. "The analogy between Mars and the earth is, perhaps, by far the greatest in the whole solar system. . . . ," he wrote. "[I]ts inhabitants probably enjoy a situation in many respects similar to ours." Herschel had no trouble believing that Mars, the other planets, and even the Sun and the Moon bore living beings. His observations further refined the Martian rotation period to 24 hours, 39 minutes—2 minutes too long—and showed that the red planet's axial tilt resembled the Earth's.

During the 1830s, German astronomers Johann Mädler and Wilhelm Beer made maps of Mars and deduced the correct rotation period, 24 hours and 37 minutes. Herschel had erred, they found, because he had missed a rotation. They also discovered that the south polar cap waxed and waned more than the north polar cap. That's because the southern seasons are more extreme than the northern seasons: Mars is nearest the Sun right before the southern hemisphere has summer and farthest right before southern winter, so southern summers are hotter and southern winters colder than their northern counterparts.

In 1858, Italian astronomer and priest Angelo Secchi said that Syrtis Major was blue: "The existence of seas and continents . . . has been conclusively proved." Noting the similarity with the ocean separating Europe and America, Secchi renamed Syrtis Major the Atlantic Canale—the first use of that fateful word on Mars. Four years later, however, Oxford geologist John Phillips said that if Mars really had seas, observers on Earth should see the Sun's reflection in them. They never did.

Of all Martian oppositions, none figured as prominently as the favorable one of 1877. When the red planet swung close to Earth, American astronomer Asaph Hall deployed the U.S. Naval Observatory's new 26-inch telescope and discovered two small moons skirting around the planet. Although Hall could discern nothing on the moons them-

The "Canals" of Mars

"That Mars is inhabited by beings of some sort or other we may consider as certain as it is uncertain what those beings may be."

—Percival Lowell, *Mars and Its Canals*

THE NEW MOONS, however, were merely a footnote to 1877's more fantastic "discovery": a network of fine lines, or canals, crisscrossing the Martian globe that suggested the existence of intelligent life. Their discoverer, Italian astronomer Giovanni Schiaparelli, had started to learn the constellations at the age of four: "Thus, as an infant, I came to know the Pleiades, the Little Wagon, the Great Wagon. . . . Also I saw the trail of a falling star; and another; and another. When I asked what they were, my father answered that this was something the Creator alone knew. Thus arose a secret and confused feeling of immense and awesome things."

Schiaparelli became director of Brera Observatory in Milan, where he earned a reputation as an excellent observer. In 1866 he was the first to link a meteor shower—the Perseids of August—with a comet, which had sailed past four years earlier. Schiaparelli found that the comet, named Swift-Tuttle, followed the same orbit as the meteoroids, leading him to deduce that every August the Earth plunged through the stream of debris the comet had shed.

Schiaparelli spied the Martian canals eleven years later, through an $8^3/_4$-inch telescope. "It is as impossible to doubt their existence, as that of the Rhine on the surface of the earth," he declared. Schiaparelli knew all about canals, for in college he

A network of fine lines suggested the existence of intelligent life

selves—they were pinpricks of light—they did reveal their master's mass. That's because Mars holds on to its moons through its gravity, whose strength depends on its mass. According to Kepler's laws, the paths and periods of the moons showed that the Martian mass was a ninth of Earth's. Thus, the planet that had first revealed Kepler's laws had its own mass unveiled through the application of those same laws.

had studied architectural and hydraulic engineering. Moreover, eight years before his discovery, the Suez Canal had opened. The Italian word Schiaparelli chose to describe the Martian features—*canali*—means either "channel" or "canal," the former a natural waterway, the latter an artificial one; Schiaparelli never said which he intended. Had the canals been built by intelligent beings? Schiaparelli's cryptic reply: it was not impossible.

At first, no one else could see the canals, even though Schiaparelli said they ran across thousands of miles of Martian land. During the next Martian opposition, Schiaparelli reported that one of the canals had doubled, which only intensified the skepticism. Finally, in 1886, French observers reported that they had confirmed the canals, although Mars was then near the far point in its elliptical orbit. Because of this confirmation, and because of Schiaparelli's reputation, even the many astronomers who saw no canals trusted they were there.

As a favorable opposition approached in 1892, Mars mania swept the world. Flamboyant French astronomer Camille Flammarion wrote that "the present inhabitation of Mars by a race superior to ours is very probable." Observers reported flashes of light from the planet, probably glints off Martian clouds or snow, but many thought they were messages from the Martians themselves. Millionaires offered rewards to those who could reply to the Martians. Some suggested carving the Pythagorean theorem into the Sahara Desert, so that the Martians would know the Earth also possessed intelligent life.

Someone even saw the Hebrew word for "the Almighty" scrawled on the Martian globe. "There can be no doubt of the observer's accuracy. . . . ," *The San Francisco Chronicle* assured readers. "True, the magnitude of the work of cutting the canals into the shape of the name of God is at first thought appalling, but there are terrestrial works which to us to-day seem no

> Had the canals been built by intelligent beings? Schiaparelli's cryptic reply: it was not impossible

less impossible. Besides, it is known that the difference in gravitation between Mars and the earth would make it easily possible to do far more work with far less energy on Mars than on the earth."

If anyone could see the Martian canals, it should have been Edward Emerson Barnard, an outstanding observer using what was then the world's best telescope, a 36-inch at California's new Lick Observatory. In 1892, Barnard had spotted the first new Jovian satellite since Galileo. "I have been watching and drawing the surface of Mars," Barnard wrote. "It is wonderfully full of detail. There is certainly no question about there being mountains and large greatly elevated plateaus. To save my soul I can't believe in the canals as Schiaparelli draws them." Barnard penned a whimsical tale about the flashes of light the Martians were supposedly sending. After some effort, Earthlings asked the Martians why they were signaling the Earth. The Martian reply: we aren't signaling *you;* we're signaling Saturn.

most pivotal—Percival Lowell. "Of all the men through history who have posed questions and proposed answers about Mars," wrote historian of science William Graves Hoyt, "the most influential and by all odds the most controversial was Percival Lowell . . . Perhaps more than any other individual, Lowell shaped the broad conceptions and misconceptions regarding Mars in particular and extraterrestrial life in general."

Born to wealth and privilege—an entire town in Massachusetts is named after the Lowells; his brother became president of Harvard University; his sister

> "I can see yet a small boy half way up a turning staircase gazing with all his soul into the evening sky."

Amy was a famous poet—Percival Lowell discovered astronomy early. "Consciously," he wrote, "I came into this world with a comet, Donati's Comet of 1858 being my earliest recollection and I can see yet a small boy half way up a turning staircase gazing with all his soul into the evening sky where the stranger stood."

When Lowell graduated from Harvard in 1876, he delivered a short talk on the solar system's origin, but for years afterward he devoted himself to the family business. In 1883 he began to explore what were then exotic lands on Earth, Japan and Korea, and wrote highly acclaimed books about them, notably *The Soul of the Far East*. A decade later, while on his final trip to Asia, Lowell reportedly heard that Schiaparelli's eyesight was failing, and he vowed to establish his own observatory to study Mars.

Unlike most astronomers of his time, Lowell had the foresight to recognize what all astronomers know today: an observatory's location is crucial to its success. "A large instrument in poor air will not begin to show what a smaller one in good air will," Lowell wrote. "When this is recognized, as it eventually will be, it will become the fashion to put up observatories where they may see rather than be seen." Yet the U.S. Naval Observatory had placed its 26-inch telescope—the one Asaph Hall had used to discover the Martian moons—in Washington, D.C., and the new Yerkes Observatory had, in Lowell's words, "buried its [40-inch] glass in Wisconsin." To seek the best site, Lowell asked an assistant, Andrew Douglass, to test the night skies at various places over the arid Arizona Territory.

Lowell chose to build his observatory in Flagstaff—population 800. Although he had missed the favorable Martian opposition in 1892, he still had time for the one in 1894. Mars was somewhat more distant, but much farther north, hence easier for American astronomers to observe. Using an 18-inch telescope that year, and a 24-inch during subsequent oppositions, Lowell studied Mars night after night, mapping hundreds of canals.

Lowell had more than a hefty purse; he also had a powerful pen. Unlike most scientists, he could write beautifully. Witness the opening of his 1895 book, *Mars*: "Once in about every fifteen years a startling visitant makes his appearance upon our midnight skies,—a great red star that rises at sunset through the haze about the eastern horizon, and then, mounting higher with the deepening night, blazes forth against the dark background of space with a splendor that outshines Sirius and rivals the giant Jupiter himself."

FACING PAGE: Percival Lowell thought he saw canals on Mars.

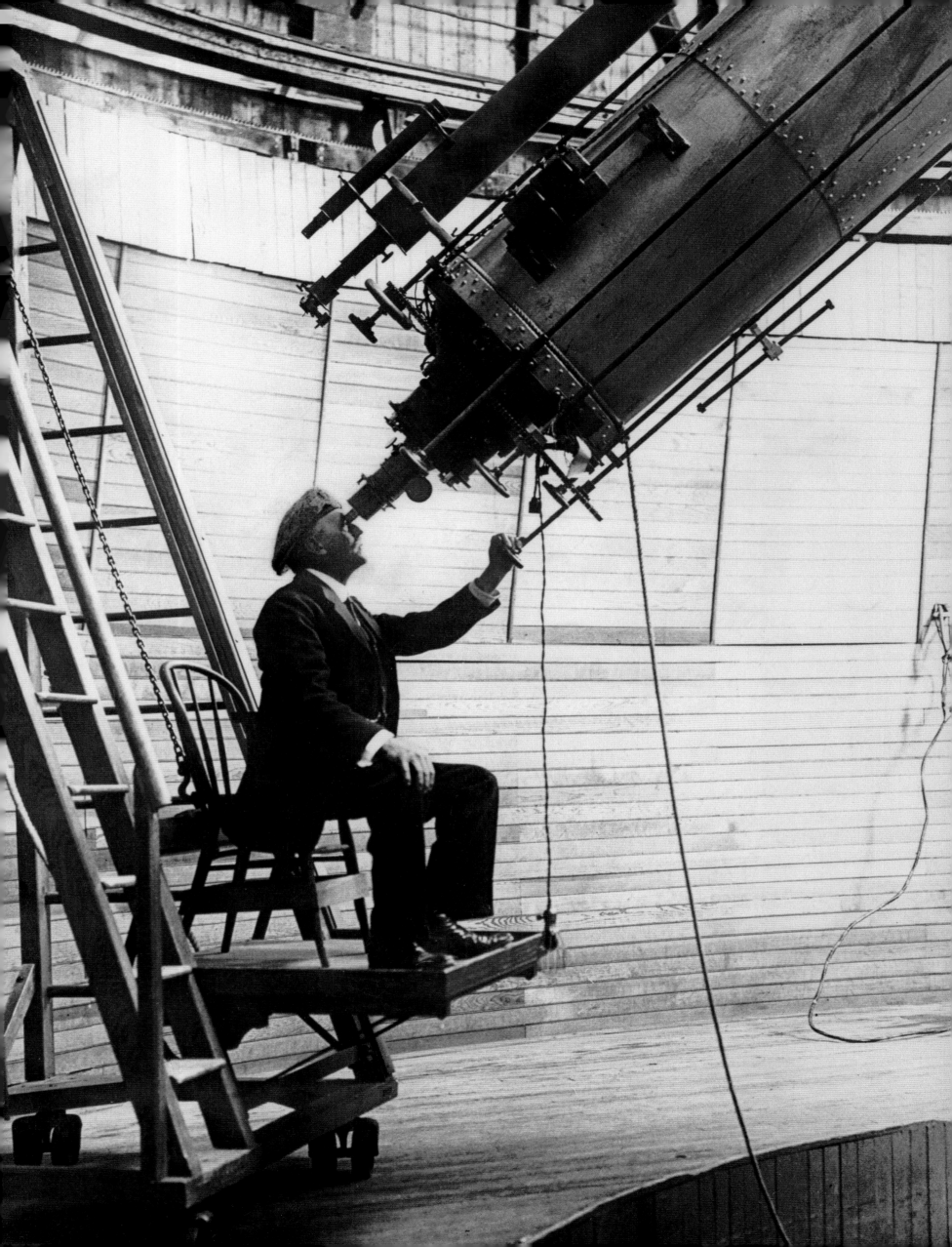

In books, articles, and lectures, Lowell depicted a wise but struggling Martian civilization that had constructed the canals to ferry precious water from the polar caps to the equator, where most of the Martians lived. Following the suggestion of William Pickering, a Harvard astronomer who had helped him establish the observatory, Lowell believed that an observed canal was really vegetation on its banks, the true waterway being too narrow for Earthbound observers to see. Lowell saw spots where the canals joined— "oases," he said, in the Martian desert. The blue-green "seas" like Syrtis Major were really vegetation, and changing colors that swept across Mars were the response of this vegetation to the seasonal influx of water from the poles. Lowell estimated that the mean Martian temperature was 48 degrees Fahrenheit, only slightly cooler than the Earth's 60 degrees.

From these observations, Lowell's theory of Martian intelligence followed almost inexorably. Straight lines are unnatural; nature prefers curves. Thus, the Martian canals had been dug by intelligent beings, to save themselves from planetary drought— for Mars was a dying, drying world. Smaller than Earth, it had evolved more quickly. It had lost its heat faster, just as a freshly baked roll cools faster than a loaf of bread. Wrote Lowell, "[T]he process that brought it to its present pass must go on to the bitter end, until the last spark of Martian life goes out. The drying up of the planet is certain to proceed until its surface can support no life at all. Slowly but surely time will snuff it out. When the last ember is thus extinguished, the planet will roll a dead world through space, its evolutionary career forever ended."

FACING PAGE: Frost blankets the floor of Lowell Crater. North is up.

> The reaction from other astronomers was severe. Most criticized both Lowell's observations and his spectacular Martian theory

The reaction from other astronomers was severe. Most criticized both Lowell's observations— the "canals," they said, were either illusions or else natural phenomena misinterpreted—and his spectacular Martian theory. Lick Observatory director Edward Holden noted sarcastically, "The conclusions reached by Mr. Lowell at the end of his work agree remarkably with the facts he set out to prove before his observatory was established at all." Holden called Lowell's writings "very misleading and unfortunate; and all the more so because they are very well written." Even Lowell's loyal assistant Andrew Douglass expressed doubt: "His method is not the scientific method and much of what he has written has done him harm rather than good," Douglass wrote in a confidential letter. "I fear it will not be possible to turn him into a scientific man." The letter got leaked to Lowell—and Lowell fired him. Douglass went on to make a name for himself by using tree rings to study past climate.

atmosphere blurs the planet's image. "I remember the excitement with which I read Lowell's [1906 book] *Mars and Its Canals,*" one astronomer later wrote. "I was about twelve then, and the book made a profound impression upon me. I believed it absolutely. In fact, it was inconceivable to me that anyone could *not* believe it." *The Wall Street Journal* called 1907's most extraordinary event "the proof afforded by the astronomical observations of the year that conscious, intelligent human life exists upon the planet Mars. . . . [I]t is by no means an impossible stretch of the imagination

> In the eyes of many, the next Martian opposition killed Lowell's colorful theory

to believe that as our mastery over electrical forces becomes more complete, we may be able before the present century ends to establish some sort of communication with the people of Mars."

The scientific establishment soon struck back. For one thing, many people saw no canals on Lowell's photographs. For another, said naturalist Alfred Russel Wallace, any canals "would be the work of a body of madmen rather than of intelligent beings," because their water would evaporate in the desert air long before it reached the thirsty Martians. Wallace

dioxide.

In the eyes of many, the next Martian opposition killed Lowell's colorful theory. The 1909 opposition was favorable, but astronomers at the world's new largest telescope—a 60-inch atop California's Mount Wilson—saw no canals. Nor did Eugène Antoniadi, who used a 33-inch telescope in France.

Lowell died seven years later, still believing in the Martian canals and the beings who had built them. He achieved a posthumous vindication of sorts in 1930, when Lowell Observatory astronomer Clyde Tombaugh discovered a faint, far-off world that Lowell himself had apparently predicted. "If it should be a planet," wrote one astronomer, reflecting the contempt many still felt for Lowell, "it is the greatest joke that ever happened to the astronomical profession: it was predicted by the amateur Lowell; the telescope was made largely by another amateur . . . ; the object was actually found by a sub-amateur assistant; and now that they have the observations there are many who suspect that no one on the Lowell staff knows how to compute the orbit." Not coincidentally, the first two letters of this planet's name—Pluto—are the initials of Percival Lowell.

During Lowell's life and long after his death, his vision of the red planet lived on through science fiction. In 1897, English author H. G. Wells wrote a terrifying story in which Martians invaded Earth. They were ultimately defeated only by the lowly terrestrial bacteria to which they had no defense. Four decades later Orson Welles transformed this story into a radio drama that scared Americans who mistook it for the real thing. Novels of Edgar Rice Burroughs depicted green-skinned, canal-tending Martians. And in 1950,

Ray Bradbury's *The Martian Chronicles* reversed things by making Earthlings the invaders of Mars and watching from afar as their peers on Earth destroy themselves in a nuclear war.

Meanwhile, in the world of science fact, a more modest Mars was rising: a cold, dry, desert world that nevertheless supported life in the form of hardy plants. During the 1920s, astronomers found that parts of the planet approached room temperature, and they reported—incorrectly, scientists now know—that Martian air possessed decent quantities of oxygen and water vapor. The following decade, however, astronomers failed to find any oxygen, and in 1947 Dutch-born American astronomer Gerard Kuiper detected carbon dioxide, then thought to be a minor component but now known to dominate the Martian air. Occasional sightings of canals even kept the hope of intelligent life alive.

"If we were to sum up in a single sentence our present knowledge of *physical* conditions on Mars," wrote French astronomer Gérard de Vaucouleurs in 1954, "we might liken them to those which would obtain in a terrestrial desert, shifted to the polar regions and lifted to stratospheric level. We leave it to the reader to decide whether under such circumstances Mars can be 'The Abode of Life' or not." As British astronomy writer Patrick Moore wrote in 1956, "There is no reason to suppose that low forms of vegetation may not exist on Mars, while there is a great deal of evidence that they do. On the other hand, the thin, oxygen-poor atmosphere is certainly unable to support either animals or men."

Spacecraft to Mars

EVEN THIS LESSER MARS, however, could not withstand the scrutiny of visiting spacecraft. The first two American planetary spacecraft, Mariners 1 and 2, had aimed for Venus, the easiest planetary target. Mariner 1 veered off course and was deliberately blown up, but Mariner 2 flew past Venus in 1962, showing that the planet was extraordinarily hot.

Mars was a tougher target. It required a longer trip farther from the Sun. Mariner 2 had derived its power from two solar panels, but sunlight is so weak near Mars that Mars-bound Mariners required four solar panels, making them look like celestial windmills.

> Mars was a tougher target. It required a longer trip farther from the Sun

On November 5, 1964—two days after Lyndon Johnson trounced Barry Goldwater in the presidential election—the first American Mars-bound spacecraft, Mariner 3, left Earth. At first all seemed well, but an hour later flight controllers were alarmed that Mariner 3 was not receiving any power from its solar panels. A shroud, meant to protect the spacecraft before and during launch, had failed to eject, preventing the spacecraft from unfurling its solar panels. Commands to shake the shroud loose also failed. As a result, Mariner 3 was forced to subsist on battery power. Eight hours and forty-three minutes after launch, the battery went dead, and the spacecraft fell forever silent.

NASA had planned to launch a twin spacecraft, but now scientists and engineers scrambled to redesign its shroud. Finally, on November 28, 1964, Mariner 4 lifted off and successfully ejected its shroud. It flew by Mars on July 14, 1965, passing 6,100 miles, or one and a half Martian diameters, from the red planet's surface. Mariner 4 had traveled 325 million miles, the longest voyage in human history. Its radio signals, zipping through space at light speed, took 12 minutes to reach the Earth. The Soviet Union had also tried to send spacecraft to Mars, but all had failed; SOVIET PROPAGANDA LOSS FROM MARINER IS NOTED, said *The New York Times*.

Goodbye to the Mars of Percival Lowell

Mariner 4 made three major discoveries. The first was actually a nondetection: no magnetic field. At most, the Martian magnetic field was 1/1000 as strong as the Earth's, so charged particles from the Sun—the solar wind—slam into the planet's atmosphere. On Earth the terrestrial magnetic field protects life from this radiation. The field itself arises from currents that circulate in the Earth's molten core. The absence of a

strong magnetic field on Mars suggested that the planet's core was no longer molten.

Mariner 4's second discovery concerned the planet's atmosphere. "If there is life on Mars," wrote Walter Sullivan on the front page of *The New York Times,* "it almost certainly does not fly, for the density of the air is too low." Mariner measured the atmospheric thickness through a clever but risky maneuver. After photographing Mars, but before sending those photographs back, the spacecraft flew behind the planet, cutting itself off from Earth for 54 minutes. If it never reestablished contact, all its photographs would be lost. By passing behind the planet, however, Mariner could beam its radio signal through the Martian atmosphere, allowing scientists to analyze the intervening air. During the 1950s, astronomers had thought the planet's surface atmospheric pressure was about 85 millibars—8.5 percent of the Earth's 1,000-millibar atmosphere. That's close to what Percival Lowell had thought. Two years before Mariner 4's flyby, however, astronomers had studied the carbon dioxide in the Martian atmosphere and claimed that the total atmospheric pressure was only around 25 millibars. Mariner 4 found that even this figure was too optimistic. The actual surface pressure was a mere 4 to 7 millibars—about 1/200 of the Earth's. The small quantity of carbon dioxide that Gerard Kuiper had earlier detected must actually make up most of the atmosphere.

Mariner 4's most shocking discovery came last, as the spacecraft sent back those photographs of the Martian surface. There were only twenty-two of them, blurry and black-and-white; each took over eight

FACING PAGE: The Odyssey spacecraft heads to Mars.

bad news. They flew closer to Mars than their predecessor and took two hundred pictures, showing ten times as much area—and, for the most part, ten times as many craters. They found that the south polar cap had the temperature expected if it was carbon dioxide ice, about −190 degrees Fahrenheit.

> No canals, no forests, no civilizations—just craters ranging from a few miles to 75 miles across

Fortunately, the next missions to Mars unveiled a far more interesting world. The previous spacecraft had merely flown past and caught fleeting glimpses—Mariner 4 photographed 1 percent of the planet, Mariners 6 and 7 together 10 percent—but Mariners 8 and 9 were to *orbit* Mars, a more difficult feat, to provide a prolonged view of nearly the whole planet. Both spacecraft were launched in May 1971. However, Mariner 8's rocket misfired, and the craft plunged into

favorable oppositions in 1909 and 1924 and the near-favorable opposition of 1941. Charles Capen, an astronomer at Lowell Observatory, even predicted a global dust storm for 1971.

Sure enough, in late September, astronomers sighted a yellow cloud hovering over Noachis, a cratered region in the south. The cloud grew and grew until it shrouded nearly the entire planet. On November 13, 1971, Mariner 9 became the first spacecraft ever to circle another planet, but it saw little more than the south polar cap and four dark spots near the equator, somehow jutting above the raging storm. The rest of the planet was a dusty blur. To conserve power, mission controllers ordered Mariner 9 to stop taking pictures and wait out the storm.

The Soviet spacecraft arrived later and also successfully went into orbit around Mars, but they were as rigid and inflexible as their country of origin. Preprogrammed to carry out their mission, both shot landers into the thick of the storm: the first lander crashed; the second landed but ceased all transmissions seconds later. The two orbiters fared little better, snapping picture after picture of a featureless globe.

Meanwhile, Mariner 9 waited. The dust began to clear, and the four dark spots near the equator proved to be giant volcanoes far taller than any mountain on Earth. Stretching away from the volcanoes was a canyon now named in the spacecraft's honor Valles

Marineris that ran across 2,500 miles—the distance from Cleveland to San Francisco. Mariner 9 also revealed that the northern half of Mars had a lower elevation than the southern and that the north was mostly smooth, the south heavily cratered. By bad luck, the previous Mariner spacecraft had photographed primarily the south, suggesting that craters blanketed the globe. Intriguingly, the southern lands featured what looked to be dried-up riverbeds meandering among all the craters. Could Mars have once been a warmer, wetter world, spawning life?

Still, the planet had no canals. Most had probably been the illusions of hopeful astronomers, although one canal did line up with Valles Marineris. The blue-green colors that suggested seas or vegetation were illusions, too, a trick of the eye: most of Mars is orange and red, so the eye perceives regions lacking these colors as their opposites, blue and green. Reported changes in these colors were probably partly the work of dust storms.

The next American mission to Mars, the ambitious Viking project, successfully placed two spacecraft into orbit and landed two more on the surface. Viking's goal was simple but difficult: to search for life on Mars. The Viking 1 lander had been scheduled to set down July 4, 1976, to commemorate the bicentennial of the American Revolution. However, photographs from the Viking 1 mothership showed that the landing area was rough and hilly, threatening to topple the lander if it came down on a boulder or steep slope. Therefore, scientists postponed the touchdown while Viking scouted out smoother pastures, away from obvious dangers, away even from craters, since the impacts which formed them would have shot out rocks too small for the orbiter to see but large enough to tip the lander over. Finally, on another famous anniversary—July 20, 1976, seven years after Neil Armstrong set foot on the Moon—the Viking 1 lander set down on the plains of Chryse, photographing a

ruddy, rocky landscape bearing no obvious signs of life: no Martians, no plants, no cars or cities or canals. The sky, originally thought to be blue, was actually pink, from airborne Martian dust. Comedian Johnny Carson joked about the discovery of the first gay planet, and indeed the discovery had been made by a gay scientist, James Pollack of NASA's Ames Research Center. His more geocentric colleagues, preferring the original blue sky, booed the announcement. The Viking 2 lander set down September 3, 1976, farther north, on the plains of Utopia, and saw a similar scene. Viking 1's latitude was 22 degrees north, similar to Honolulu's; Viking 2's was 48 degrees north, the same as Seattle's.

> The blue-green colors
> that suggested seas or
> vegetation were a trick
> of the eye

Both landers looked for life in the Martian dirt. They extended their mechanical arms, retrieved soil, added water and nutrients, and waited for life to bloom. Surprisingly, the soil emitted gases that terrestrial organisms would have, but Viking could detect no organic molecules, even though it had a sensitivity down to a few parts per billion. In the view of most but not all scientists, Viking failed to find life on Mars.

Following the Viking mission, NASA did not launch another Mars probe until the early 1990s, the expensive but ill-fated Mars Observer. In 1993, as it

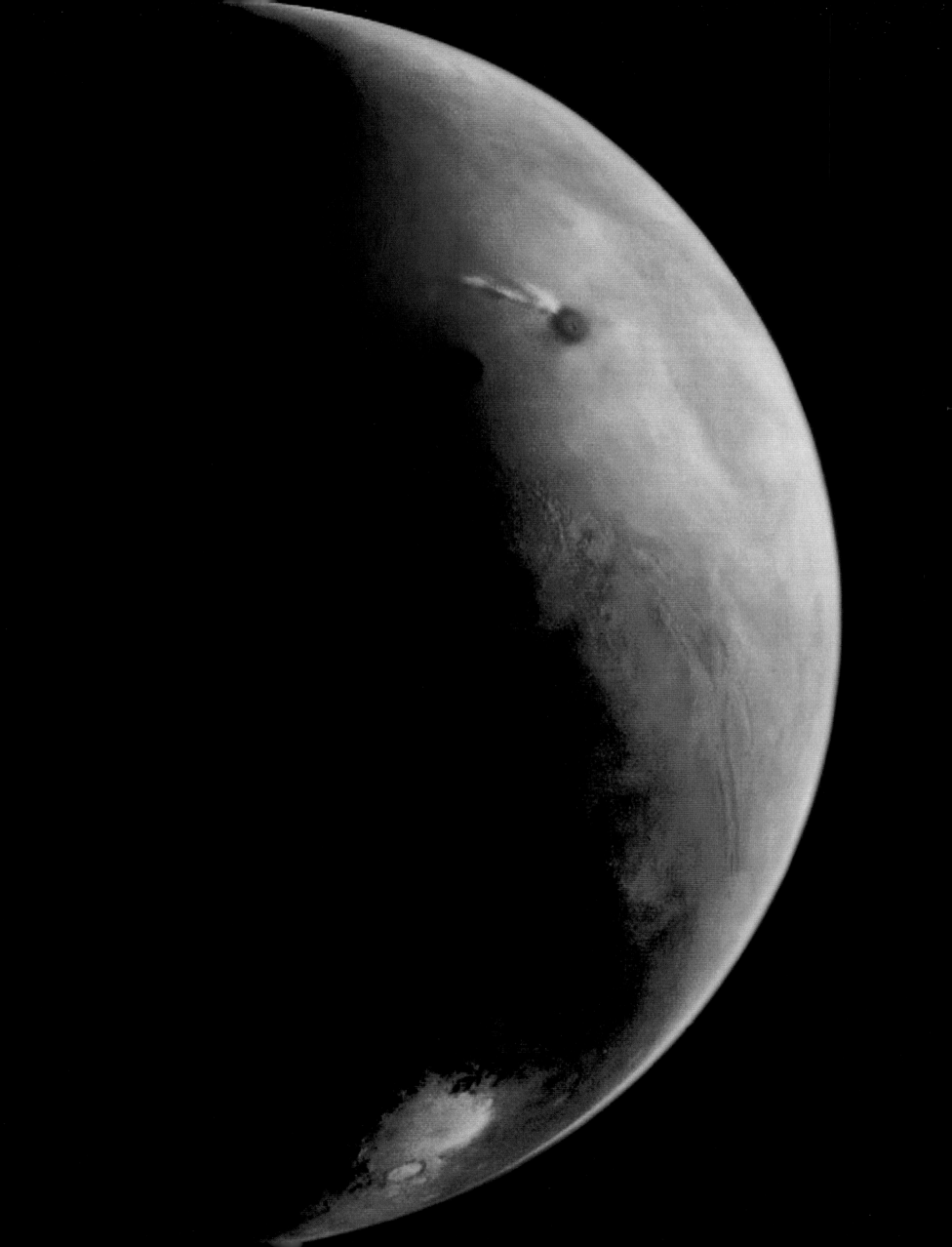

neared Mars, its fuel line apparently ruptured, sending the spacecraft spinning. It never regained contact with the Earth.

Mars next starred in the news three years later, when a piece of the planet landed on front pages of newspapers. Scientists already knew that some meteorites came from Mars, because their gases matched those Viking had detected in the Martian atmosphere. In 1996 scientists claimed that one Martian meteorite held microscopic fossils of tiny life, suggesting that the red planet had once harbored microbes. However, other scientists attacked this claim, and the controversy may not be resolved until astronauts travel to Mars and search for fossils themselves.

The following year, 1997, marked two triumphs for Mars exploration. In a spectacular touchdown on July 4, the Pathfinder spacecraft landed near Viking 1 and released a small rover that examined the Martian rocks. The mission was cheap, part of NASA's new "faster, better, cheaper" philosophy, and cost only one dollar per American. Rather than be released from an orbiter, as the Viking landers had, Pathfinder headed straight for the surface, encasing itself in a huge "beach ball"; it hit the Martian surface at 40 miles per hour, then bounced as high as a five-story building. It bounced over a dozen times before coming to rest. It landed at the mouth of Ares Vallis, a flood channel that had served up a smorgasbord of rocks, some of which, Pathfinder determined, may have formed in water. The temperature at the Pathfinder site ranged from +10 degrees Fahrenheit during the day to −100 degrees Fahrenheit just before dawn.

The Pathfinder mission proved enormously popular with the public. Within thirty days of the landing, the Pathfinder web site scored 566 million hits. Strangely, according to New Scientist, the publicity displeased NASA, which ordered the Pathfinder press

office shut down—because the spacecraft had upstaged NASA's expensive space shuttle, whose flight at the same time received almost no notice.

In contrast to Pathfinder, another Mars-bound spacecraft that year received so little attention that some scientists compared it to the radar-defying Stealth aircraft. The Mars Global Surveyor went into orbit and eventually began taking extremely detailed photographs of the surface below. Three years after its arrival, for example, scientists announced the discovery of gullies that may have been carved by recent flows of water.

Unfortunately, the next two missions to Mars failed as badly as the previous two had succeeded. Set to orbit Mars in 1999, the Mars Climate Orbiter instead burned up in the planet's atmosphere. An embarrassing mix-up between English and metric units had caused the spacecraft to veer 60 miles—um, that's 100 kilometers—too close to Mars. If the metric system had never been invented, the accident would not have occurred; but pro-metric forces managed to berate Americans for failing to go metric. Meanwhile, a second American spacecraft was heading for Mars. The Mars Polar Lander was supposed to fire two penetrators into the Martian surface and then land near the south pole, but for reasons still mysterious it crashed instead.

Following these two disasters, the next Mars mission was a success. In 2001 the Mars Odyssey spacecraft went into orbit around the planet and in 2002 discovered large quantities of ice beneath the surface. This and other spacecraft have returned images and data that now give us the best picture yet of the red planet. It's not the vibrant Mars of Percival Lowell, nor the boring Mars of Mariner 4, but a complex planet whose history and mystery scientists are just now beginning to decipher by exploring the planet's four elements—EARTH, AIR, FIRE, and WATER.

FACING PAGE: A Martian crescent greets the Viking spacecraft. North is approximately up.

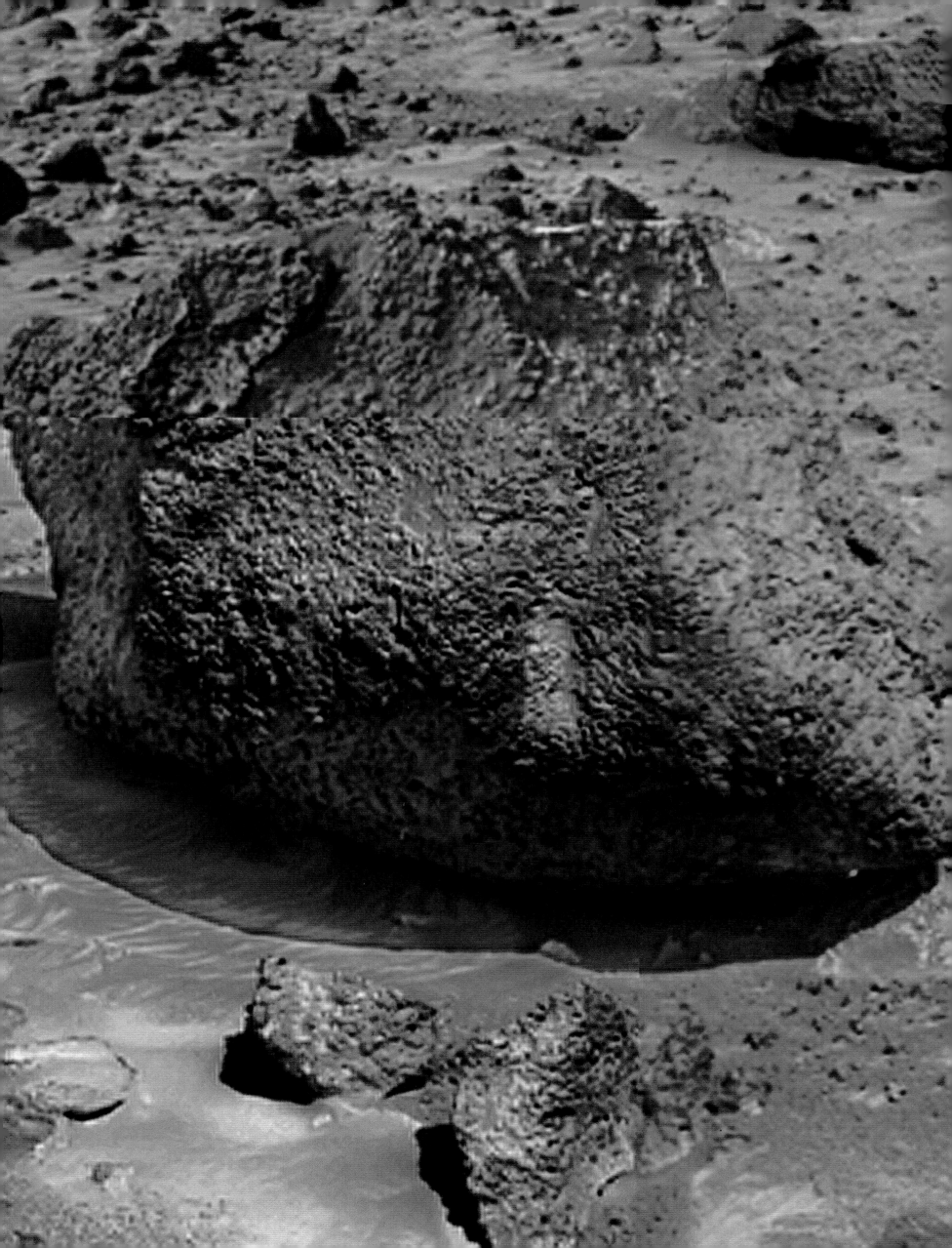

Earth

THE RUDDY MARTIAN SURFACE bears witness to 4.6 billion years of planetary history. Mighty mountains loom larger than any on Earth, monstrous marsquakes have ripped open gargantuan canyons, and wayward asteroids have punctured the Martian globe, creating gigantic craters. The canyons and craters—the scars of Mars—may have once hosted lakes larger than Earth's; dried-up riverbeds and a possible vanished ocean testify to a warmer, wetter past. Orbiting spacecraft have scrutinized the planet from above, landers have sampled the soil below, and asteroids hitting the planet have catapulted Martian rocks into space, where they later struck the Earth as meteorites. By unraveling the clues preserved in the Martian interior, surface, and meteorites, scientists hope to piece together the story of Mars—and determine whether its lakes and rivers once sprouted life whose fossils await discovery today.

FACING PAGE: Scientists named this three-foot-tall rock at the Pathfinder landing site "Yogi." The rock was not available for comment.

The Three Ages of Mars

TO DECIPHER the symphony of Martian history—a seeming cacophony of magma oceans, volcanic eruptions, asteroid impacts, torrential floods, global dust storms—scientists begin with time. As someone once said, time is nature's way of keeping everything from happening at once. Time converts a chord into a melody. It can arrange the events tangled together on the Martian surface into a story, a sequence of cause and effect. This story carries a moral, of a good planet gone bad, and thus a warning to residents of its vibrant neighbor. Exploring Mars, then, might help save the Earth.

In many ways, time is easier to study there than here. That's because the Earth has largely torn up its birth certificate. The same terrestrial activity that sustains life—continental drift, volcanic eruptions, winding rivers, rainfall—obliterates the Earth's past. As the continents grind across the globe, they trigger volcanic eruptions that recycle carbon dioxide, a greenhouse gas that warms the world and thereby helps life; but this continental drift also erases the ocean floor, so geologists don't even know the continents' positions a billion years ago. The volcanoes' lava buries old terrain, while rain, snow, and wind eat away at it. Thus, old Earth rocks are rare. Although the Earth's age is 4.6 billion years, the most ancient known terrestrial rocks, in the Northwest Territories of Canada, are 4.0 billion years old, and the oldest known fossils, in Western Australia, 3.5 billion years old. Scientists already have a rock from Mars older than both.

That's because Mars has better preserved its youth. It has no continental drift, its volcanoes rarely erupt, rain never falls, and the air is thin. Therefore, much of the planet's surface still documents the era when life was struggling to establish itself on Earth, and possibly on Mars. Although the Earth teems with life, its dead red neighbor may better reveal how that life first arose.

On both worlds, time manifests itself in the layers of rock and soil the eons have laid down. By carving Arizona's Grand Canyon, the Colorado River exposed strata hundreds of millions of years old, the youngest on top, the oldest on bottom. Through stratigraphy—the study of such strata—geologists have assigned periods like the Tertiary, Cretaceous, and Jurassic to the Earth's past.

Likewise, stratigraphy frames Martian history. If a volcano spews lava over an impact crater, the lava flow must be younger than the crater; if a river then slices through the lava, the resulting riverbed must be younger than both the lava and the crater. People with messy desks employ the same stratigraphic principle. To retrieve a recent document, they look near the top of the mess; to find an older one, they look beneath. Another technique for dating Martian terrain exploits the planet's many craters. The older a surface, the more meteorites have hit it, so the more battered it is. In the same way, the older a dartboard, the more holes it has.

Following these methods, scientists recognize three Martian periods. Conveniently, when arranged from youngest to oldest, they appear in alphabetical order: the Amazonian period, the Hesperian period, and the Noachian period (mnemonic: **Amaz**ing **He**roes **No**d). The youngest period, encompassing all recent activity, is the Amazonian, named for a smooth plain in the northern hemisphere of Mars. The Amazonian

FACING PAGE: Lying west of the volcano Olympus Mons, the smooth plains of Amazonis Planitia lend their name to the most recent period of Martian history. North is up.

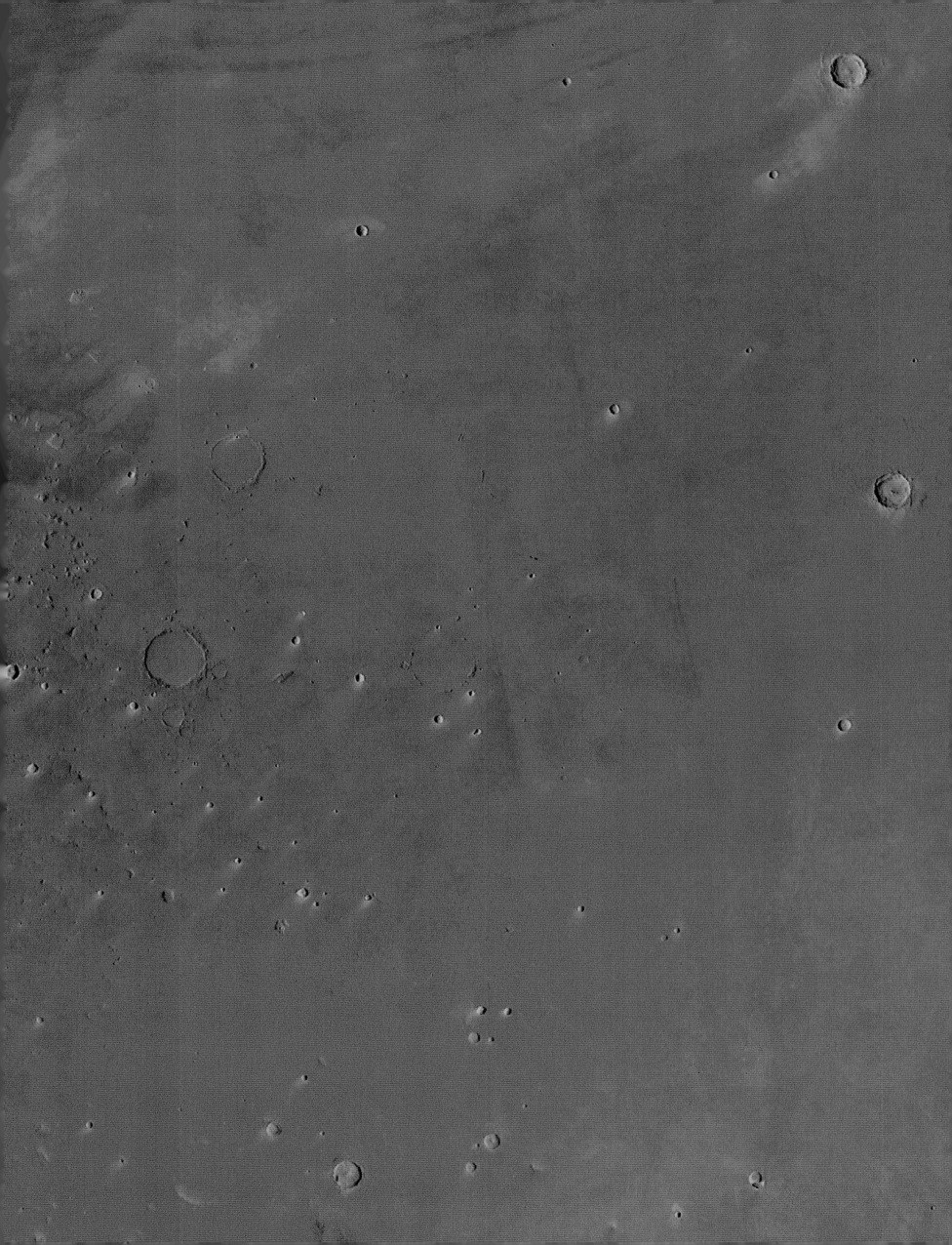

period accounts for 26 percent of Martian territory, including much of the northern hemisphere as well as the ice caps at both poles. The next youngest period, the Hesperian, spans the Martian middle ages and takes its name from a moderately cratered plateau in the southern hemisphere. Terrain of Hesperian age covers 34 percent of the planet. The oldest Martian period, the Noachian, features the most heavily scarred territory, especially much of the southern highlands. Noachian terrain blankets 40 percent of Mars. This period is named for a heavily cratered region in the far south of Mars, which itself is named for Noah, the Old Testament figure who built an ark prior to a flood. Appropriately, Noachian-era Mars sported rivers, possible lakes, and perhaps an ocean. Even before spacecraft glimpsed the red planet, "Noachian" meant ancient and antiquated. For further precision, scientists divide each period into subperiods. The Amazonian period has three subperiods:

How long did each of the three Martian periods last?

the Early Amazonian, the Middle Amazonian, and the Late Amazonian; the Hesperian period has two subperiods: the Early Hesperian and the Late Hesperian; and the Noachian period has three subperiods: the Early Noachian, the Middle Noachian, and the Late Noachian.

How long did each of the three Martian periods last? Unfortunately, no one knows. On Earth geologists date strata by exploiting how radioactive elements in rocks decay into daughter elements. The older the rock, the fewer radioactive elements endure and the more daughter elements accumulate. The

ratio of the first to the second, plus knowledge of the radioactive element's half-life, yields the rock's age. In this way geologists know that the Earth's Cretaceous period gave way to the Tertiary period 65 million years ago, the same time an asteroid hit the Earth and killed off the dinosaurs.

However, none of the spacecraft that touched down on Mars dated the rocks strewn before it, nor has any Martian spacecraft returned rocks to Earth for analysis. Scientists *have* dated meteorites from Mars, but without knowing where on Mars the rocks came from, whether from Amazonian, Hesperian, or Noachian territory, scientists can't use these ages to quantify the stratigraphic record.

Instead, scientists extrapolate from the Moon. The Moon has two main terrains: the smooth, dark lunar seas, such as the Sea of Tranquillity, and the light-colored cratered highlands. By dating the rocks the Apollo astronauts retrieved, scientists learned that during the first 800 million years of its life the Moon and presumably the Earth suffered heavy bombardment from asteroids and comets. That's why the lunar highlands are so heavily cratered. After the heavy bombardment ceased, lava flooded the low-lying lunar seas and erased craters, so the lunar seas bear few scars today.

The heavy bombardment presumably occurred throughout the solar system, so Mars also suffered numerous impacts from 3.8 to 4.6 billion years ago. Then the impact rate plummeted. In trying to date the

FACING PAGE: Hesperia Planum, a moderately cratered plateau south of the equator, typifies the middle period of Martian history. North is up.

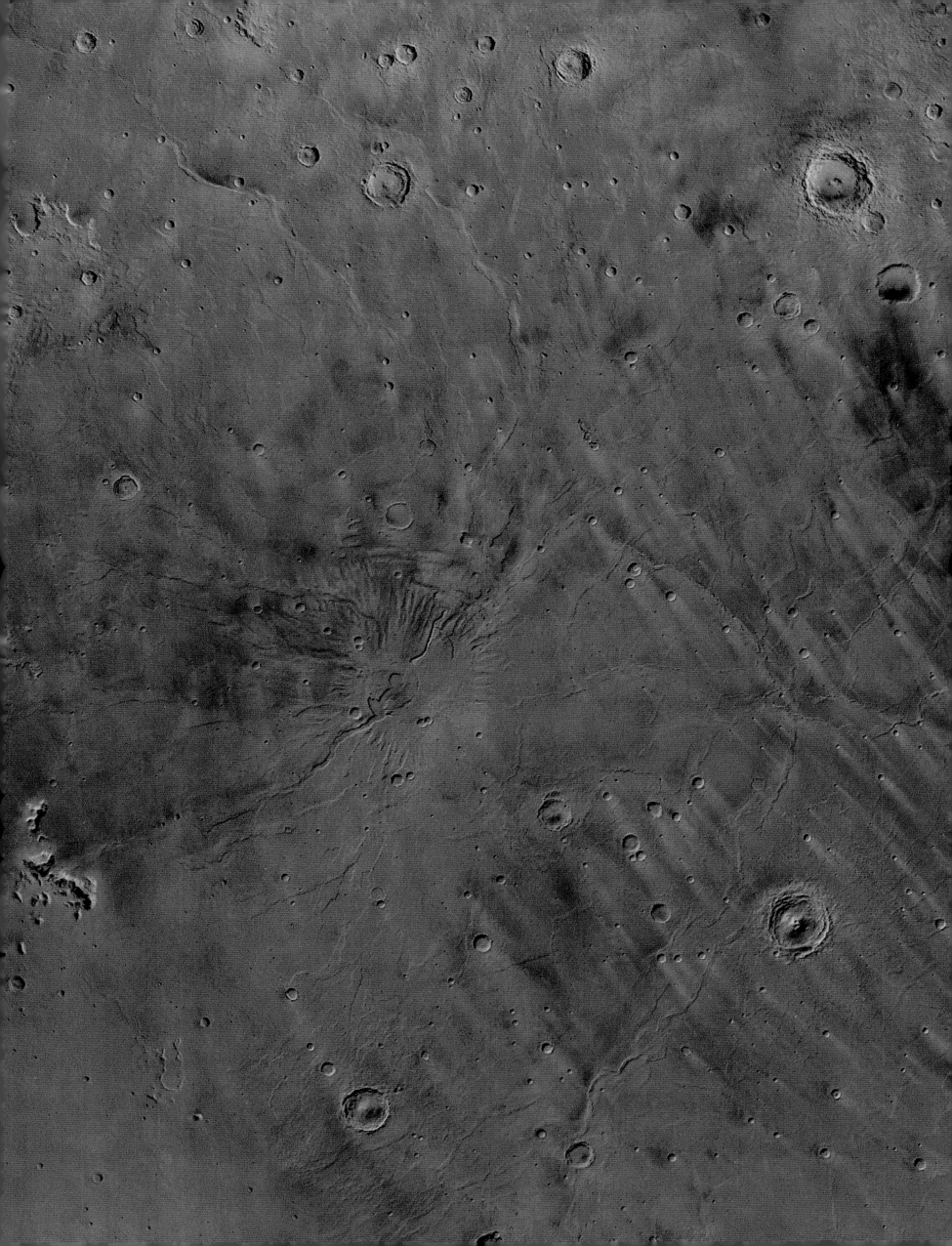

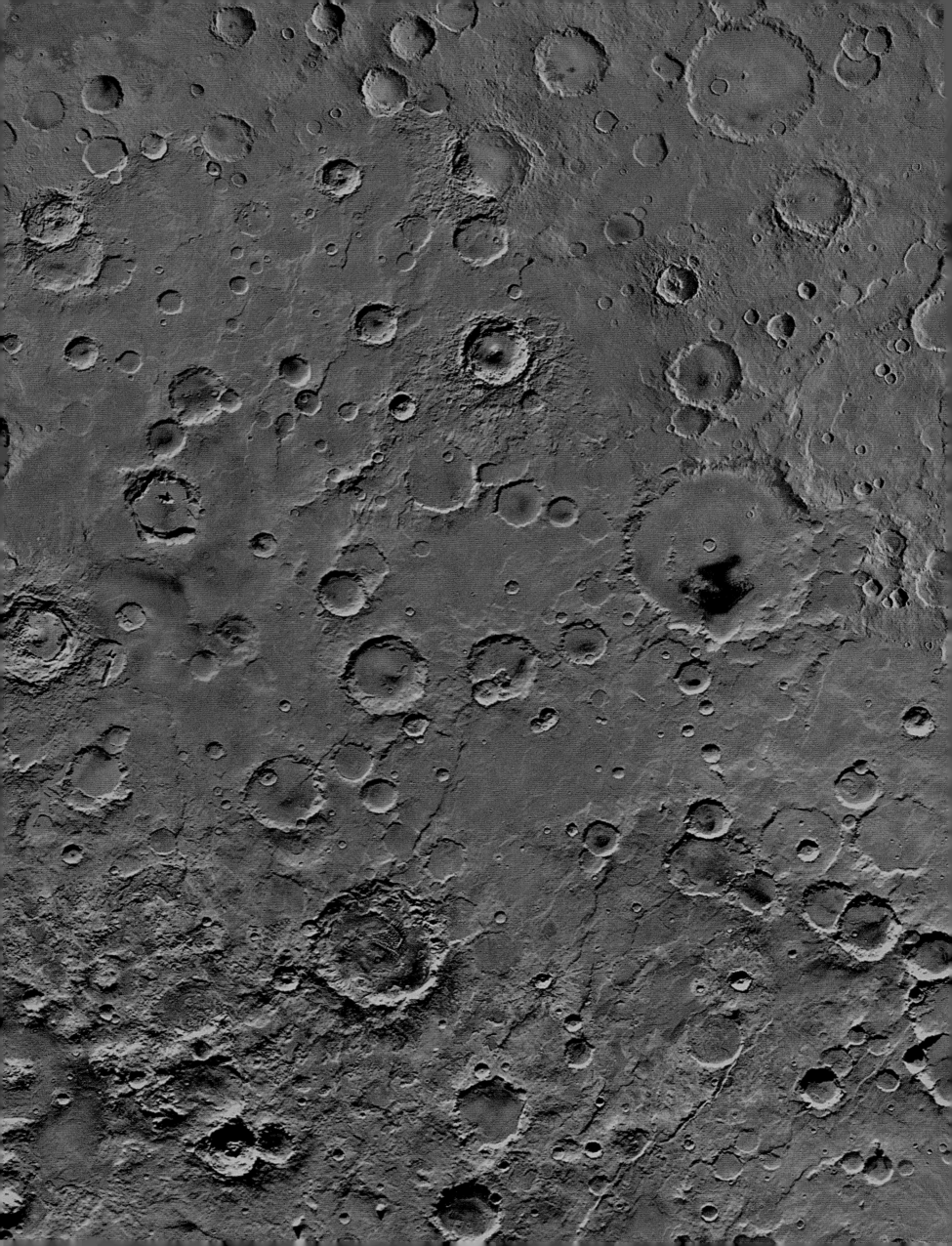

Martian surface by applying the lunar record, however, scientists face several difficulties. First, Mars probably suffers more impacts per area per year than the Moon—it resides closer to the asteroid belt—so a Martian terrain bearing the same crater density as a lunar terrain is younger. Current estimates suggest that Mars suffers about twice as many impacts per area per year as the Moon, and that asteroid impacts far outnumber comet impacts. Second, the thicker atmosphere surrounding ancient Mars prevented some impacts by incinerating many would-be impactors and then eroding the craters that still managed to form. Also, dust from the dust storms buries small craters.

The uncertainties in the cratering rate obscure the dates of the three Martian periods. In 2001, William K. Hartmann of the Planetary Science Institute in Tucson, Arizona, and Gerhard Neukum of Deutsches Zentrum für Luft- und Raumfahrt in Berlin, Germany, published work that tried to narrow widespread differences between previous studies. By analyzing the impact rate on Mars, they concluded that the Amazonian period began roughly 3.1 billion years ago, which means this period likely spans the most recent two thirds of Martian history. The Late Amazonian, they estimated, began a mere 300 to 600 million years ago. The Hesperian period extended from about 3.1 to 3.6 billion years ago, making it the shortest of the three Martian periods. And the Noachian period stretched from about 3.6 to 4.6 billion years ago, when the solar system itself formed, so the Noachian period included both the heavy bombardment and its immediate aftermath.

FACING PAGE: The battered highlands of Noachis Terra, in far southern Mars, give their name to the oldest period of Martian history. North is up.

The Birth of Mars

LIKE ITS PLANETARY brothers and sisters, Mars arose in a disk of gas and dust that spun around the newborn Sun. Nearest the Sun the disk spun fast, so friction raised the temperature and only tough substances with high melting points, such as iron and silicates, condensed and formed solid particles. These particles then conglomerated into asteroid-like objects called planetesimals, which smashed together to build Mercury, Venus, Earth, and Mars. Hence, the four inner planets consist mostly of iron and silicate rock.

> Mars arose in a disk of gas and dust that spun around the newborn Sun

Farther from the Sun, beyond the asteroid belt, the disk of gas and dust spun more slowly, so it was colder, allowing ice to condense. Because ice was so common and the outer disk so large, the outer planets—Jupiter, Saturn, Uranus, and Neptune—grew into giants. Indeed, the gravity of Jupiter and Saturn attracted huge amounts of the disk's two most common elements, hydrogen and helium, so these two planets swelled into gas giants. In contrast, the inner planets were too small for their gravity to hold on to such light elements.

Leftover debris—asteroids, comets, far-off Pluto—lingered. For 800 million years, during the heavy bombardment, asteroids and comets pummeled the planets. Then, for reasons unknown, the storm stopped. Perhaps the Sun and its planets finally swept up most of the interplanetary flotsam, or perhaps some cataclysm ejected it from the solar system.

As the solar system's largest planets, Jupiter and Saturn also helped end the onslaught. Together the two gas giants harbored over twelve times the mass of all the other planets combined, and their gravity tossed trillions of comets out of the solar system. In 1992, George Wetherill of the Carnegie Institution of Washington suggested that if Jupiter and Saturn had not arisen, neither would we. Because the two gas giants ejected so many comets, he said, few of these deadly bodies hit the Earth today. Devastating impacts occur roughly every 100 million years—the last was 65 million years ago—giving advanced life time to evolve. Great comets akin to Halley, Hyakutake, and Hale-Bopp pass Earth about once a decade. Without Jupiter and Saturn, though, devastating impacts would occur so often—about once every 100,000 years—that they would impede the evolution of advanced life. If stargazers still managed to arise, they'd be treated to the sight of a hundred great comets a year.

Although Jupiter's birth was good for the Earth, it was bad for Mars. As Prussian philosopher Immanuel Kant recognized in 1755, Jupiter's great gravity stirred up the planetesimals trying to build Mars, stunting the red planet's growth. Wrote Kant, "It is probably the neighborhood of the very large planet Jupiter which, by attraction on its side, has robbed Mars of the particles needed for its formation." Without Jupiter, Mars would have sprouted into a larger world, perhaps comparable to the Earth in mass; but Jupiter's interference cut the actual Mars down to a world with 89.3 percent less mass than the Earth. The planet's present problems—frigid temperature, thin air, lack of volcanism and continental drift—stem in part from its small size. Thus, without Jupiter, Mars might have blossomed into a second Earth. Jupiter did even worse to the planet that tried to develop between itself and Mars: an asteroid belt marks the ruins of a stillborn planet.

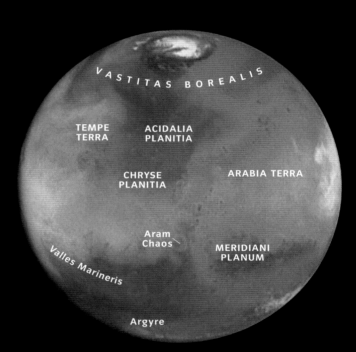

ABOVE AND FACING PAGE: As Mars turns: this and the next five images depict a full rotation of Mars. In all images, north is up, and the north polar cap is at top; the south polar cap does not appear, because it was tilted away from Earth.

In this view, centered around longitude 20 degrees west, most of the large dark patch covering the north-central region is Acidalia Planitia; the Viking 1 and Pathfinder landers set down to its southwest, in Chryse Planitia. Below and left of center are the canyons of Valles Marineris, and the white patch near bottom is the Argyre impact basin.

Prior to the 1980s, scientists thought that Mars had been born cold. According to this now-discredited idea, radioactive elements inside Mars decayed and slowly heated the planet's interior until it partially melted. At that point, perhaps a billion years after the planet's birth, the iron sank through the lighter silicate rock and congregated at the planet's center. The creation of this dense iron core further heated Mars, just as water spilling over a dam releases energy.

In the 1980s the meteorites from Mars revealed instead that the planet had been born hot, from the heat the planetesimals released when they smashed into Mars, likely giving the young world an ocean of magma. As a result, almost as soon as Mars was born, the iron slipped through the silicate rock and formed the core, supplying more heat to the young world.

Remarkably, the Martian meteorites date the iron core's formation to within just 13 million years of the red planet's own birth—well less than 1 percent of the planet's present age. In 2002 scientists in Germany led by Thorsten Kleine of the University of Münster reached this conclusion by studying the heavy metals hafnium and tungsten, atomic numbers 72 and 74. Both are tough, like iron, with high melting points, but they differ in their affinity for iron. Hafnium dislikes iron, so when the iron core formed, the hafnium stayed behind, in the silicate mantle and crust. In contrast, tungsten is a siderophile ("iron-loving") element, so it slipped out of the mantle and crust when the iron did. As an alliterative mnemonic, think of the hafnium halting in the crust, while the tungsten got tugged out of it.

The Martian meteorites come from the crust, so most have more hafnium than tungsten. Nevertheless, they have an unusual amount of one tungsten isotope, tungsten-182. That's because it can arise when radioactive *hafnium*-182 decays. Hafnium-182 decays into tungsten-182 with a half-life of just 9 million years. If the iron core had formed slowly—say, over a billion years—then nearly all the hafnium-182 in the crust would have turned into tungsten-182 and then gotten tugged into the core when the iron finally sank. Because the Martian core actually formed fast, however, much of the tungsten-to-be was still in the form of hafnium; this halted in the crust, and only later decayed into tungsten-182, accounting for its high abundance in the Martian meteorites. Thus, the core formed during the very earliest years of the Early Noachian period.

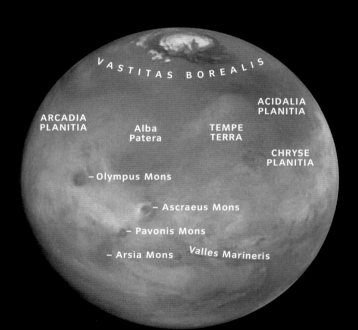

ABOVE AND FACING PAGE: In this view, centered on longitude 94 degrees west, Acidalia Planitia has rotated to the upper right; Valles Marineris is the greenish line running east-west on the lower right. The mighty volcano Olympus Mons appears at the left center, an orange peak poking through white clouds; to its southeast appear three more volcanoes: Arsia Mons, Pavonis Mons, and Ascraeus Mons.

Global Properties

EACH PLANET in the inner solar system— Mercury, Venus, Earth, Mars—has an iron core encased in a mantle and crust of silicate rock. However, the proportions of iron and rock vary from planet to planet. Mercury has a large iron core, Mars only a small one. The first clue to a planet's composition is its density—mass divided by volume. If you've ever lifted a birthday gift to deduce its nature, you've used the same method: a dense gift might be a hardcover book, a light one a piece of clothing.

To calculate the Martian density, scientists must measure the Martian mass and the Martian diameter. Astronomers have known the mass ever since they tracked the planet's moons, which feel its gravity. Later, the trajectories of Mars-bound spacecraft refined the measurement. Mars has only 10.7 percent as much mass as the Earth. As a result, if a satellite circled as far from Mars as the Moon does from the Earth, it would require not one but nearly three months to revolve.

The Martian diameter comes from telescopic and spacecraft observations. In 1999 the Mars Global Surveyor's laser altimeter found the equatorial diameter nearly a mile smaller than previously thought, 6,792.4 kilometers or 4,220.6 miles. Mars has about half the diameter of the Earth and twice the diameter of the Moon. Of the solar system's nine planets, the Earth ranks number five in size—four planets are larger, four are smaller—and Mars ranks number seven.

Dividing the Martian mass by its volume reveals that the planet's mean density is 3.94 times that of water. The Earth is denser, 5.52 times water's density, and the Moon lighter, at 3.34. Iron is over twice as dense as rock, so the Earth must have a greater proportion of iron than Mars, and the Moon a lesser proportion.

People visiting Mars may want to walk slowly. In 1998 scientists simulated Martian gravity aboard an airplane and found that the best walking speed, which minimizes the amount of energy consumed per mile, is 2.1 miles per hour. The best walking speed on Earth is 60 percent faster, 3.4 miles per hour. The good news for future Martians is that the lower gravity means walking a mile will require only about half as much energy as it does on Earth.

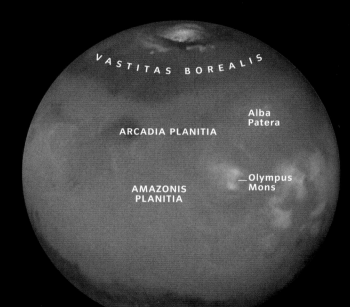

ABOVE AND FACING PAGE: In this view, centered on longitude 160 degrees west, Olympus Mons has rotated rightward—it's the white patch right of center. The small white patch to its northeast is the peculiar volcano Alba Patera, and to its southeast the three volcanoes Arsia Mons, Pavonis Mons, and Ascraeus Mons still appear. The featureless orange region at center is Amazonis Planitia; the orange region above it, containing two dark patches, is Arcadia Planitia. The large dark collar above Arcadia is Vastitas Borealis. On the left, the volcanoes of Elysium are rotating into view.

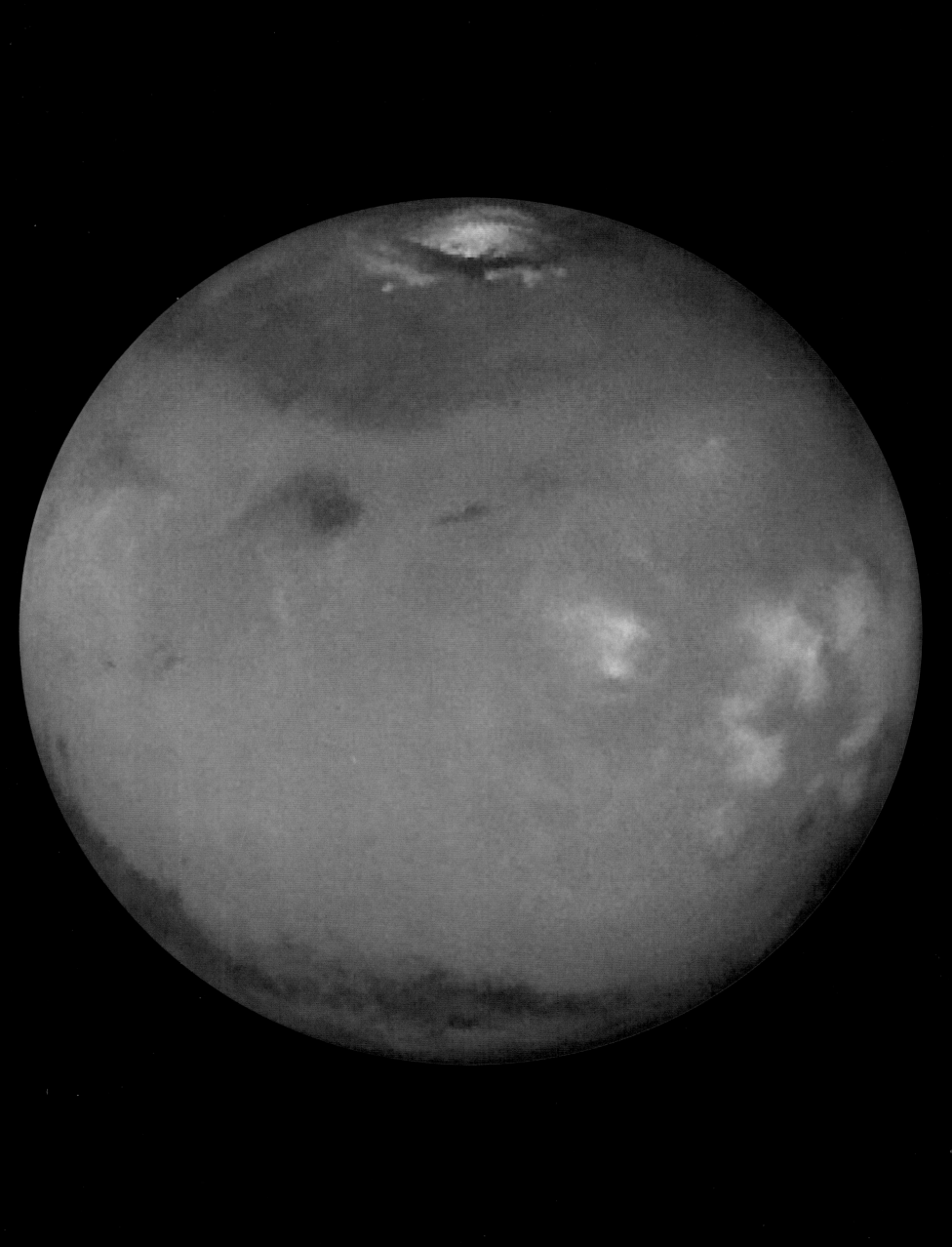

Martians will experience days similar to Earth's, because the red planet rotates every 24 hours and 37 minutes. This period is not the day Martians should set their clocks to, however, because as Mars spins it also moves a bit around the Sun. Due to this movement, the Sun does not occupy exactly the same place in the sky after the planet completes a full rotation. Instead, the planet has to rotate a bit more for the Sun to catch up. As a result, the time between successive noons on Mars is slightly longer—24 hours and 40 minutes. In the same way, the Earth rotates every 23 hours and 56 minutes, but Earthlings think a day is really 24 hours, because of the Earth's movement around the Sun. The first period is called the sidereal period, because it's what an observer watching the planet from the stars would see—*sidus* is Latin for "star"—and the second period, which the Martians would experience, is called the synodic period.

Although the near coincidence between the two planets' rotation periods makes Mars seem more Earthlike, it is only a modern phenomenon. The Earth was born spinning faster, but over billions of years the Moon's gravity slowed it down. The Moon will continue to slow the Earth's spin. Someday, a terrestrial day will last longer than a Martian day. In contrast, the Martian moons are so tiny they barely affect their master, so the red planet's present rotation rate must be close to the one it received following all the impacts that created it.

As the Earth spins, its north pole points at Polaris, the North Star, a second-magnitude star at the end of the Little Dipper's handle less than 1 degree from the Earth's north celestial pole. Martians will be less fortunate, for the bright star nearest its north celestial pole is Deneb, the brightest star in Cygnus the Swan, but it's 9 degrees away. So Mars has no good North Star. However, unlike the Earth, it does have a decent *South* Star: Kappa Velorum, in the constellation Vela the Sails. It shines 3 degrees from the Martian south celestial pole and is somewhat fainter than Polaris.

Martians will see a brilliant celestial object that their neighbors on Earth don't—the Earth itself. It would look much as Venus does in the Earth's sky, a beautiful evening star in the west or a morning star in the east; but it would boast a faithful companion, the Moon. At its best, the Earth would appear about as bright as Venus does in Earth's sky and climb about as high.

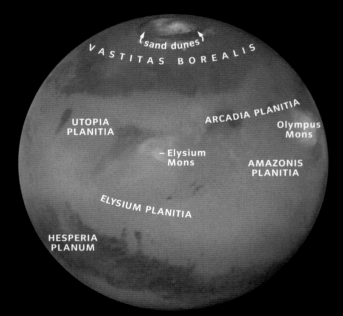

ABOVE AND FACING PAGE: Centered on longitude 210 degrees west, this view has carried Olympus Mons to the right limb. The large, mostly dark region covering much of the far north is Vastitas Borealis; the tight dark chain around the north polar cap consists of sand dunes. The bright spot near the center is the volcano Elysium Mons. To its northwest is Utopia Planitia, where the Viking 2 lander set down. Hesperia Planum appears at lower left.

Mars Inside and Out

LIKE THE EARTH, Mars has a core, mantle, and crust. The Earth's core is mostly iron. Other likely ingredients include oxygen, nickel, and sulfur. Sulfur is a crucial element for the Martian core, because it can help keep it liquid. Mars might have more sulfur than Earth, since sulfur has a fairly low melting point and so should have condensed more readily at Mars's distance from the Sun.

Unfortunately, investigating the Martian core is difficult. Earthquakes probe the Earth's interior, but the Viking 1 lander's seismometer failed to work, and Viking 2's detected no marsquakes. So scientists try to model the Martian interior by knowing its density and its moment of inertia, or how spread out its mass is, which scientists inferred in 1997 by comparing Pathfinder data with Viking data to determine how much the planet wobbles as it rotates. According to these models, if the core is pure iron (atomic number 26), it is small and dense; if the core also contains a lighter element, such as sulfur (atomic number 16) or hydrogen (atomic number 1), it is larger but less dense.

A key clue to a planet's core comes from its magnetic field, because on Earth this arises from currents circulating in the liquid part of the core stirred up by heat. Thus, if a planet has a magnetic field, its core must be at least partially molten so that heat can flow the way it does in a boiling pot of water. Ever since 1965, when the Mariner 4 spacecraft flew past the red planet, scientists have known that Mars has no substantial magnetic field. Soviet spacecraft later incorrectly suggested otherwise, but the magnetometers on NASA's Mars Global Surveyor confirmed Mariner's original finding. By 2001 they had determined the strength of any global magnetic field to be less than 1/100,000 that of the Earth's. Thus, explorers can leave their compasses at home.

However, in its youth, Mars *did* have a magnetic field. The Mars Global Surveyor's magnetometers picked up strong magnetization over part of the planet's ancient southern hemisphere. The magnetization was coming not from the core but from the crust. If a crustal rock bearing the right minerals is molten and then cools in a magnetic field, its atoms align with the field and the rock becomes magnetized. The right minerals include magnetite, pyrrhotite, hematite, maghemite, and titanomagnetite, all of which contain iron.

ABOVE AND FACING PAGE: This view, centered on longitude 288 degrees west, shows the dark triangular region Syrtis Major at center. The white patch near the right limb is the volcano Elysium Mons, and the white patch at bottom is the Hellas impact basin.

The crustal magnetization is especially intense over Terra Cimmeria, named for a seafaring people that Homer mentioned in *The Odyssey,* and Terra Sirenum, named for the sirens whose sweet voices snared passing sailors. Both Terra Cimmeria and Terra Sirenum are ancient, heavily cratered regions dating back to the Early Noachian period. Their strongest magnetizations outdo the strongest on Earth, near the Russian city of Kursk, twenty times over, and they stretch over a thousand miles, far larger than any on Earth. More than 4 billion years ago, then, Mars must have had a strong global magnetic field. This field did not last long, though, because the two large impact basins in the southern hemisphere—Hellas and Argyre—are not magnetized. Since the impacts that created Hellas and Argyre melted rock which then cooled but did not become magnetized, the Martian magnetic field had by then already decayed. The Hellas and Argyre impactors hit Mars during the Early Noachian period, so the magnetic field survived for only a few hundred million years.

The absence of a magnetic field might mean that the Martian core has solidified. Surprisingly, however, scientists in 2003 reported just the opposite: the core is at least partially liquid. They deduced the core's state by observing how the planet responds to tidal tugs from the Sun. The firmer the core, the less it should flex; but the orbit of Mars Global Surveyor, which changes with the changing shape of Mars, suggests instead that the core is partially or entirely liquid. If so, the absence of a magnetic field implies little heat flow within the core, because a vigorous heat flow would churn the liquid and create a magnetic field. Perhaps the Martian crust—which is thicker than Earth's—serves as a lid that suppresses the heat flow.

The bulk of the planet is not the iron core but the surrounding silicate mantle. Silicate rocks are those based on the elements silicon and oxygen. The mantle also bears radioactive potassium, thorium, and uranium. As these elements decay, they heat the mantle. This radiogenic heat could sustain Martian volcanism long after the heat of birth had faded, because potassium-40, thorium-232, and uranium-238 all have half-lifes of over a billion years.

Above the mantle is the crust, the topmost layer. The Earth's crust is thin beneath oceans and thicker beneath continents. The Martian crust is thinnest beneath impact basins, such as Hellas and

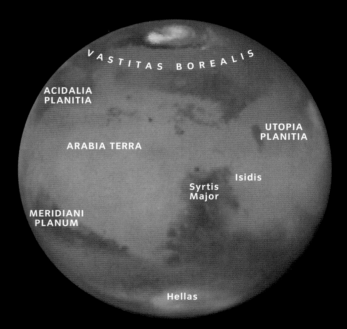

ABOVE AND FACING PAGE: Centered on longitude 305 degrees west, this image carries Syrtis Major right of center. The large orange region to its west is Arabia Terra. Meridiani Planum, through which runs 0 degrees longitude, is on the far left. What looks like the south polar cap is really the Hellas impact basin.

Argyre, and thickest beneath the volcanic province of Tharsis, presumably because the volcanoes have carried lava up from the mantle and spilled it onto the crust. The Martian crust is almost three times thicker than the terrestrial—models based on Mars Global Surveyor data suggest a mean thickness of 30 miles—because Mars has cooled and solidified more.

In part because thin crust slides more easily than thick, Earth has continental drift but Mars does not. Therefore, Mars lacks the mountain chains that arise when continents collide. Mighty though Martian mountains are, they do not line up in ranges like the Himalayas, which formed when India hit the rest of Asia, crumpling and lifting the land between.

Ancient Mars, however, may have been a different story. The crust was thinner then, aiding continental mobility, and water may have lubricated the continental plates, as it probably does on Earth. Furthermore, in 1998 the Mars Global Surveyor's magnetometers discovered possible evidence for ancient continental drift. On Earth, as the Americas drift away from Europe and Africa, lava oozes onto the central Atlantic seafloor. As this lava cools, it gets magnetized in the same direction as the Earth's magnetic field. However, the magnetic field reverses, typically after a few hundred thousand years; a compass that points north today will point south after the next magnetic reversal. Therefore, the seafloor that oozed out just before the last reversal bears the opposite magnetic polarity from seafloor forming today. Because of these reversals and the spreading of the seafloor, the Atlantic seafloor is a magnetic zebra, consisting of strips of alternating magnetic polarity.

When Mars Global Surveyor's magnetometers detected magnetized rocks in Terra Cimmeria and continental drift may explain them. But if Mars once had continental drift, and if continental drift requires water for lubrication, then ancient Mars may have sported oceans.

Martian peaks are tall and basins deep. In 1999 the Mars Global Surveyor's laser altimeter obtained crisp views of the planet's topography. The altimeter shot laser pulses that hit the surface ten times a second. The longer they took to bounce back, the lower the terrain they were hitting. The laser altimeter measured the elevation of the Martian surface to an accuracy of just a few feet. As a result, scientists now know the topography of Mars better than some parts of Earth. Furthermore, the new data indicate that old measurements were off by up to several miles. For example, the highest point on Mars—the volcano Olympus Mons—turned out to be some three miles shorter than had been thought. A similar reduction of the Rocky Mountains would put even its tallest peaks below sea level.

Mars exhibits a far greater topographic range than Earth. From the highest peak, atop Olympus Mons, to the deepest trough, in the Hellas impact basin, Mars spans an altitude range of over 18 miles. That's 50 percent *greater* than the Earth's on a world nearly 50 percent *smaller*. Why? Without continental drift, Martian volcanoes like Olympus Mons forever sit over the same hot spot in the mantle, growing taller and taller, and the thick Martian crust supports their immense weight. Furthermore, without rain or snow, peaks and valleys face little threat from erosion.

A North-to-South Tour of Mars

IT'S TIME to take a hike across Mars, from the north pole to the south, armed with the best maps of the planet. The new topographic maps, derived from the Mars Global Surveyor's laser altimeter, are as informative as they are colorful. Purples depict the planet's depths, blues the lowlands, greens and yellows somewhat higher elevations, oranges and reds the highlands, and rare touches of brown and white the tallest peaks and plateaus. Scientists define zero elevation—"sea level" on sealess Mars—as the approximate elevation the terrain would have if they smoothed out the peaks and valleys but preserved the planet's slightly nonspherical shape. More exactly, they require that the force of gravity at every point of sea level be equally strong. That's necessary because the topographic variations affect the surface gravity: a mountain like Olympus Mons exerts so much gravity that an ocean surrounding it would be lifted up—by about a mile—compared with its level were the mountain not there. It is this "lifted up" level that scientists count as zero elevation, or Martian sea level.

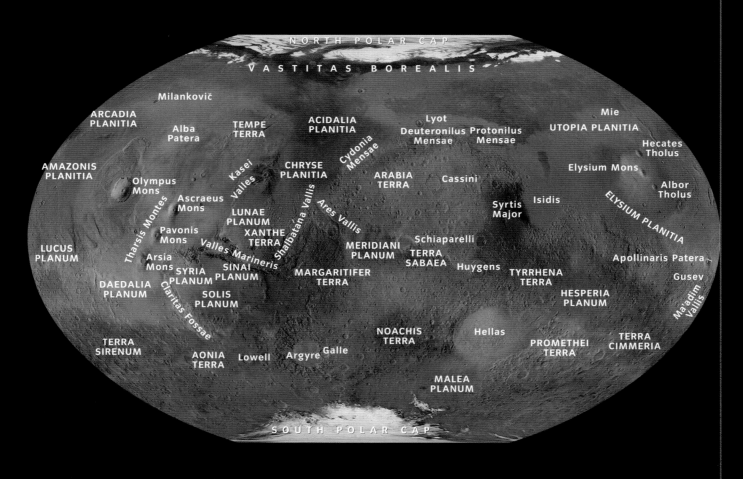

ABOVE AND PAGES 48-49: This map, centered on longitude 0 degrees, shows all of Mars. North is up.

PAGES 50-51: Full Martian topography, from pole to pole. North is up.

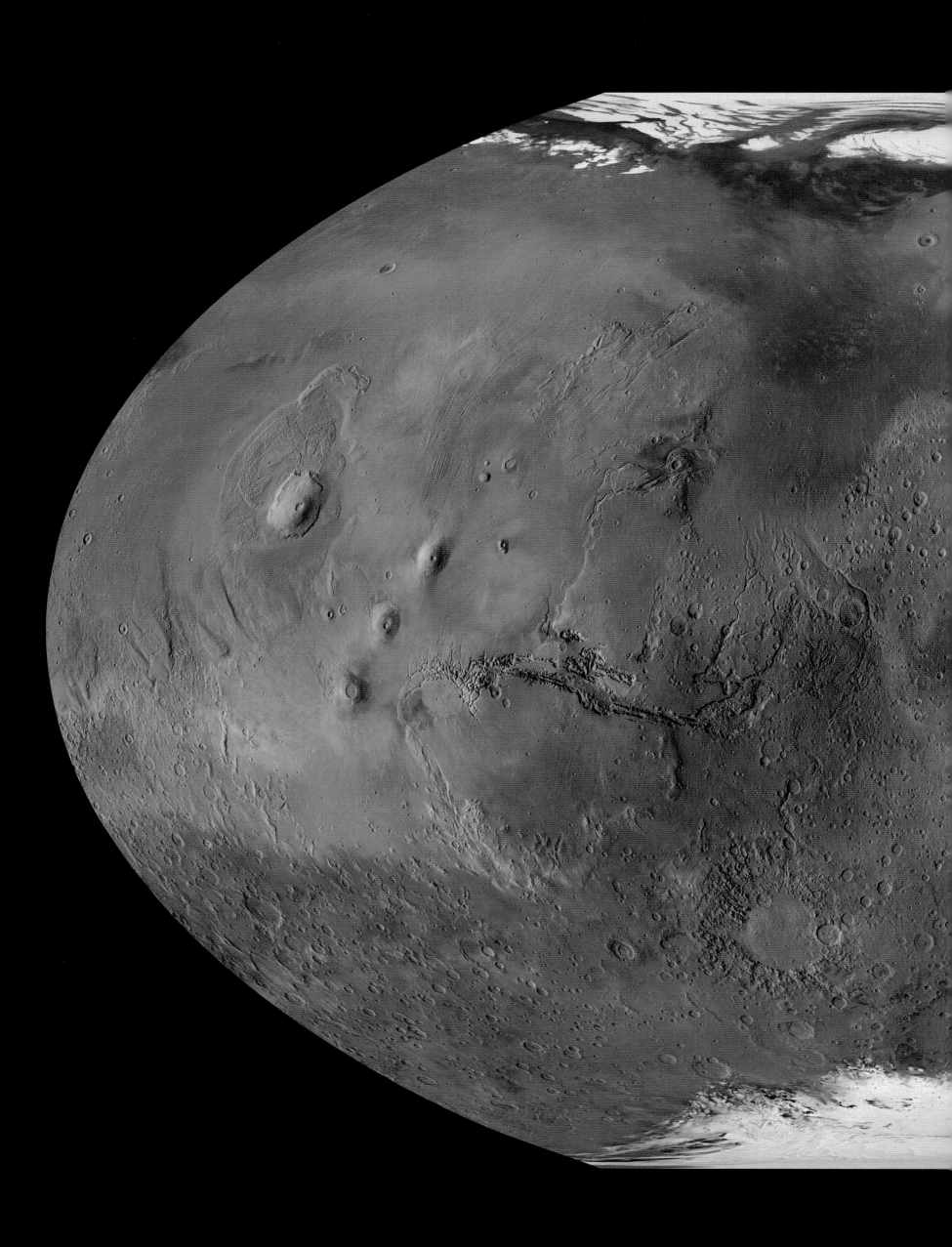

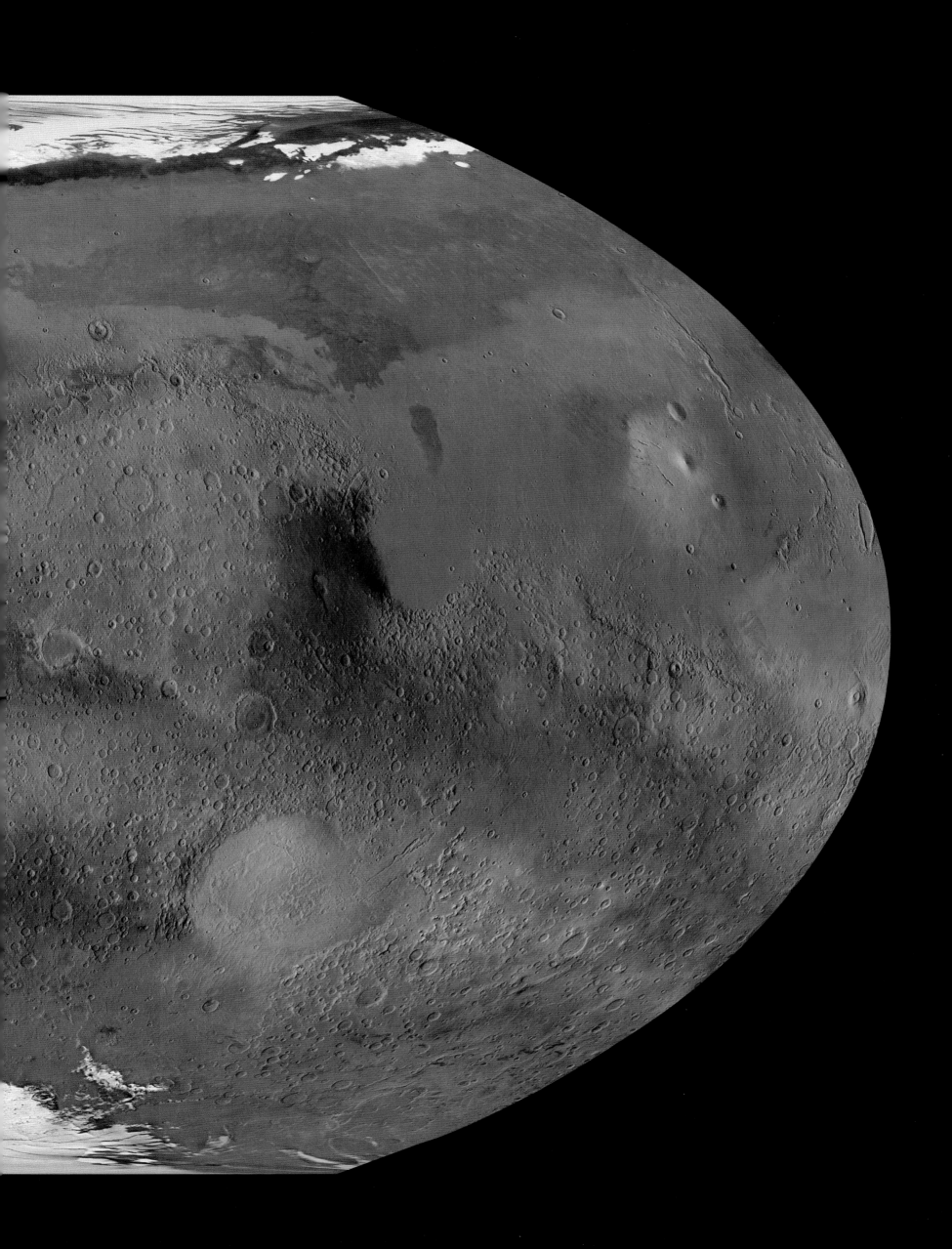

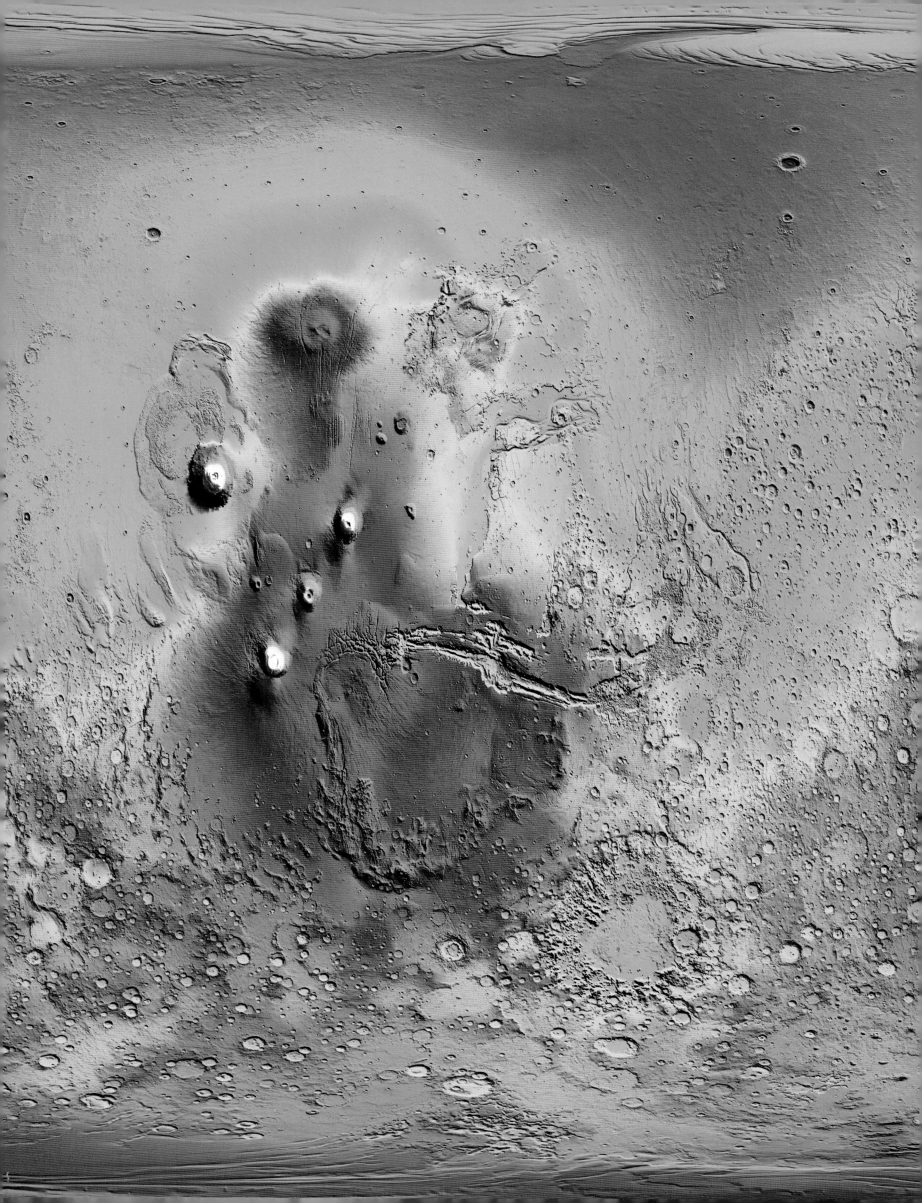

The topographic maps lay bare the startling dichotomy between northern and southern Mars. Most of the northern hemisphere, painted topographic blue, lies low; the southern, bursting with topographic orange and red, stands high. Especially striking is the color contrast between the poles: the north polar terrain is blue and green, the south mostly yellow, orange, and red. Scientists have known of this dichotomy since 1972, when the Mariner 9 mission distinguished the smooth northern lowlands from the cratered southern highlands. According to the new

> Explorers journeying north to south would start at the icy north pole. The polar cap is really two caps in one

topographic data, the north pole sits nearly 4 miles, or 6 kilometers, below the south pole, so explorers hiking from north to south would face an uphill climb. If water poured all over Mars, three fourths would drain into the northern plains.

No one knows why this north-south dichotomy exists. Perhaps the northern hemisphere happened to suffer several giant impacts that excavated rock. Perhaps the southern hemisphere happened to sit above a rising plume of hot material in the mantle. Or perhaps long ago, if the Martian continents once

drifted, they all happened to drift south, opening a basin in the north akin to that on Earth beneath the Pacific Ocean.

Explorers journeying north to south would start at the icy north pole. The polar cap is really two caps in one. The first, the residual cap, is water ice; it survives the summer heat. The second, the seasonal cap, appears during fall and winter, when carbon dioxide frost settles onto the residual cap and extends beyond it. Not surprisingly, the polar cap bears few craters. It is young, of Late Amazonian age. Surrounding the polar cap is a ring of dark sand dunes, also young, as well as a peculiar layered terrain that probably reflects past climates on Mars, discussed in AIR.

Low-lying plains—the expanses of topographic blue—dominate the rest of northern Mars. The northern plains are smooth, like the lunar seas, and bear few obvious craters. However, recent images from Mars Global Surveyor suggest abundant craters just beneath the surface, as if a thin veneer of material buried them. On the topographic map the uniform color of the northern plains denotes a uniform elevation, suggesting to some that Mars may have possessed a northern ocean, further discussed in WATER.

In 2003, Michael Carr of the U.S. Geological Survey and James Head III of Brown University published work that examined the history of the northern plains. Using images from the Mars Global Surveyor, they found that the northern plains consist of thick lava dating back to the Early Hesperian period, buried beneath about 100 meters of sediments laid down during the *Late* Hesperian—the same age as several flood channels that feed into the northern plains. Therefore, suggest Carr and Head, these Late Hesperian floods carried water into the northern plains, possibly creating a temporary ocean, which laid down the sediments.

Whichever way explorers proceed from the north pole, they cross the far northern plain called Vastitas Borealis, which stretches down to 50 or 60 degrees north latitude. Its latitude is similar to Canada's. Its name is appropriate: *vastitas* means "an extensive plain," and *borealis* means "northern," just as the aurora borealis means the northern lights.

Voyaging south of Vastitas Borealis, explorers reach the other northern plains. Their names bear the appendage *planitia,* Latin for "plain," and on Mars the term further means a low plain. Three northern plains start with an *A.* Strangely, the name of one, Acidalia Planitia, means Venus. Southwest of Acidalia is Chryse Planitia, the northern plain where the Viking 1 and Pathfinder spacecraft set down. The other two *A* plains are on the other side of the planet: Arcadia Planitia, named for a rural part of Greece—"arcadian" means idyllically pastoral—and to its south Amazonis

FOLLOWING PAGES:

Left: The low-lying north polar region is mostly topographic blue.

Right: The south polar region consists of cratered highlands, colored topographic yellow and orange.

Gatefold: The Pathfinder spacecraft captured a dramatic 360-degree panorama of the Martian surface.

Planitia, named for the Amazons, which in turn lent its name to the most recent period of Martian history. West of Arcadia is the northern plain Utopia Planitia, where the Viking 2 lander set down; southwest of Utopia is another northern plain, Isidis Planitia. Both Isidis and Utopia are impact basins, depressions carved out by impacting asteroids. Because Isidis is circular, scientists had recognized its nature in

> Utopia might be an impact basin but this was confirmed only with the topographic map

the 1970s. In the 1980s they proposed that Utopia might be an impact basin, too, but this was confirmed only with the topographic map shown here—note the nearly circular region of purplish blue that blankets Utopia.

Jutting between Utopia and Arcadia/Amazonis is the small volcanic province of Elysium, a lonely island of yellow, orange, and red in a sea of topographic green and blue. Elysium sports three volcanoes, the tallest being Elysium Mons—*mons* means "mountain." Lava from the Elysium volcanoes spills into Elysium Planitia to the south, the green plain near the equator.

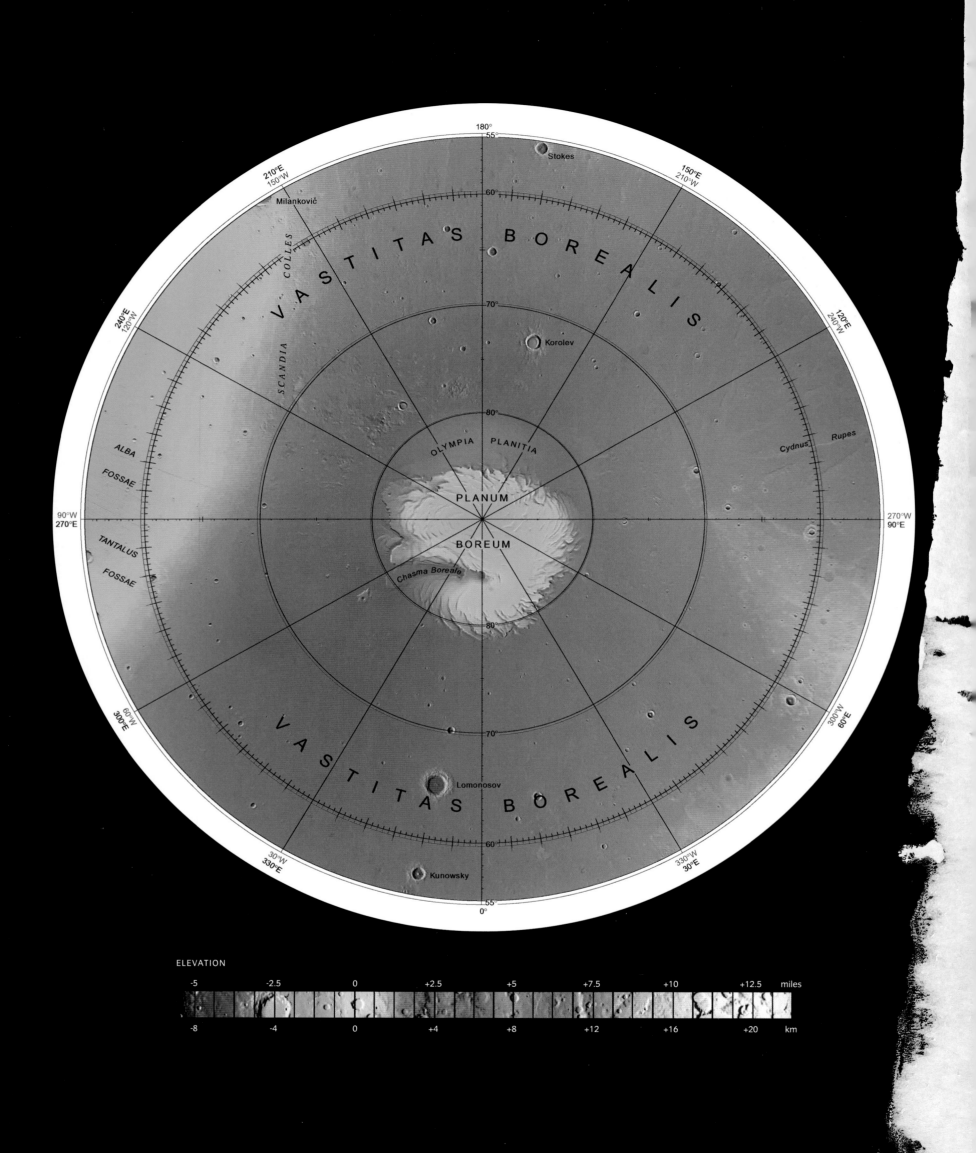

180°
55°

Stokes

Milanković

V. COLLES

SCANDIA

V A S T I T A S   B O R E A L I S

60°

70°

Korolev

80°

OLYMPIA   PLANITIA

Cydnus   Rupes

ALBA

FOSSAE

PLANUM

BOREUM

TANTALUS

FOSSAE

Chasma Boreale

80°

70°

Lomonosov

V A S T I T A S   B O R E A L I S

60°

Kunowsky

55°
0°

ELEVATION

| -5 | | -2.5 | | 0 | | +2.5 | | +5 | | +7.5 | | +10 | | +12.5 | miles |

| -8 | | -4 | | 0 | | +4 | | +8 | | +12 | | +16 | | +20 | km |

NOACHIS

Russell
Darwin
Chalcoporous
Rupes
SISYPHI
Wegener
Maraldi
Daly
Lyell
Sisyphi Cavi
PLANUM*
Sisyphi Montes
Pityusa
Rupes
Peneus
Patera
Phillips
Mellish
TERRA
Pityusa
Patera*
Malea
Patera*
MALEA
Amphitrites
Patera
Dana
Barnard
Axius
Von Karman
South
Valles
Du Toit
Joly
Dorsa
Argentea
Main
Dorsa
PLANUM
Mad
Vallis
Argyre Rupes
Fontana
PLANUM*
Schmidt
CAVI
ANGUSTI
Holmes
Mitchel
AONIA
AONIA
ARGENTEA
PLANUM
ANGUSTI
PROMETHEI PALNUM*
Promethei Rupes
Vishniac
Gilbert
PLANUM*
AONIA
Agassiz
PLANUM
Chasma Australe
PROMETHEI
Coblentz
Heaviside
AUSTRALE
Lials
Hutton
Huxley
Bianchini
PARVA PLANUM*
Rayleigh
Secchi
Smith
Lau
Promethei
Burroughs
Weinbaum
TERRA
Ross
Steno
Thyles
Heinlein
TERRA
Lamont
Chamberlin
Byrd
Rupes
Wells
Stoney
Reynolds
Dokuchaev
Richardson
Jeans
Uxis Rupes
CARIA
Charlier
PLANUM
CHRONIUM
Eridania
Scopulus
Clark
Suess
TERRA
Trumpler
FOSSAE
Keeler
TERRA
CIMMERIA
Wright
SIRENUM
Mendel
Kuiper

ELEVATION

-5   -2.5   0   +2.5   +5   +7.5   +10   +12.5   miles

-8   -4   0   +4   +8   +12   +16   +20   km

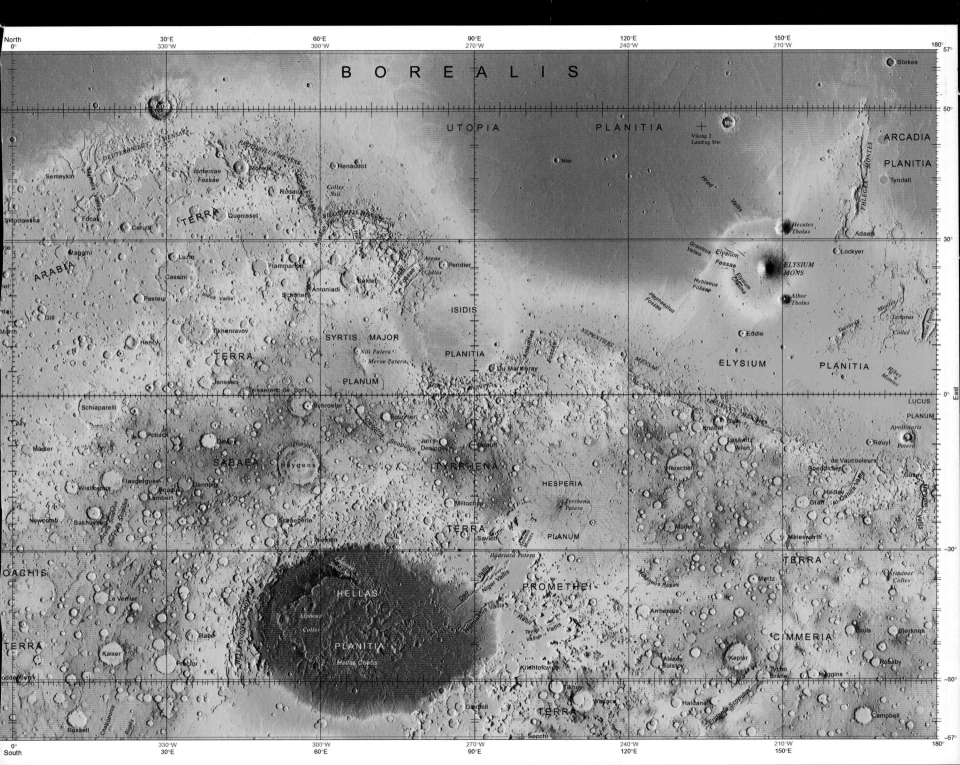

The most impressive volcanic province, though, is on the other side of the planet. The Tharsis bulge colors nearly half of Mars brilliant shades of topographic orange and red. On and around Tharsis stand five giant volcanoes: Olympus Mons is the white and brown peak northwest of Tharsis proper; Alba Patera is the circular reddish patch to its northeast; and Arsia Mons, Pavonis Mons, and Ascraeus Mons are the three white peaks that form a diagonal line straddling the equator—together these three are called the Tharsis Montes, *montes* being the plural of *mons*. The Tharsis bulge resembles movie stars who look younger than their age. It formed during the Noachian period, but lava has obliterated impact craters, so its

Dried-up riverbeds cut through the cratered highlands, ancient craters are heavily eroded, and most craters smaller than

Billions of years ago,
when Mars was warmer
and wetter, both Hellas
and Argyre may
have housed lakes

smaller than ten miles across have been washed out altogether. Thus, the Noachian period had a much higher erosion rate, suggesting, as do the ancient rivers themselves, that Mars once had a thicker atmosphere and wetter surface.

The southern highlands sport two large impact basins, Hellas and Argyre. Hellas bears the old name for Greece. To telescopic observers Hellas had looked bright, and they thought it a plateau. In fact, Hellas is the deepest impact basin in the solar system, reaching depths of over 5 miles, or 8 kilometers, below Martian sea level—deep blue and purple on the topographic map. During winter, frost forms on its floor, making it look bright. Hellas is 1,400 miles, or 2,300 kilometers, across. In size and shape, it resembles the Moon's South Pole-Aitken Basin. When the asteroid hit Mars to create Hellas, it excavated enormous quantities of rock and catapulted it throughout the south of Mars, partially explaining why the rest of the south stands so high. The impact fractured the surrounding territory, allowing magma to reach the surface; four volcanoes appear on Hellas's northeastern and southwestern rims.

The Argyre impact basin is smaller and shallower than Hellas, its floor topographic green and green-blue. Like Hellas, Argyre looks bright—its name comes from a legendary island of silver at the mouth of

the Ganges River—when winter frost blankets its floor. Billions of years ago, when Mars was warmer and wetter, both Hellas and Argyre may have housed lakes.

Between Hellas and Argyre is cratered Noachis Terra, namesake of the most ancient period of Martian history. However, the very oldest territory on Mars may reside on the other side of Hellas—Terra Cimmeria and Terra Sirenum—since these have the strongest magnetization, dating back to the era when Mars still had a magnetic field.

Explorers reaching southernmost Mars would end their adventure as they began, with a polar ice cap. Unlike its northern counterpart, the south polar cap's carbon dioxide frost lingers year long, so observers never see the water ice beneath. Indeed, scientists once debated whether the south polar cap even had water ice. The new topographic data argue for its presence, because carbon dioxide ice is too weak to support the heights the polar cap attains.

FACING PAGE: Centered on the Utopia impact basin, which is purplish blue, this topographic sphere shows the spiral-shaped north polar cap at top and the three Elysium volcanoes, colored red, at lower right.

FOLLOWING PAGES:

Page 58: This topographic sphere shows the same scene from the south. The purple scar is the impact basin Hellas, surrounded by orange and red highlands. The three red Elysium volcanoes appear near the limb at the two o'clock position.

Page 59: The Tharsis uplift colors much of this topographic sphere bright red; the white peaks are all volcanoes. The blue canyons of Valles Marineris slice east-west through Tharsis. The north polar cap appears at top.

ELEVATION

-5		-2.5		0		+2.5		+5		+7.5 miles
-8		-4		0		+4		+8		+12 km

ELEVATION

| -5 | -2.5 | 0 | +2.5 | +5 | +7.5 miles |
| -8 | -4 | 0 | +4 | +8 | +12 km |

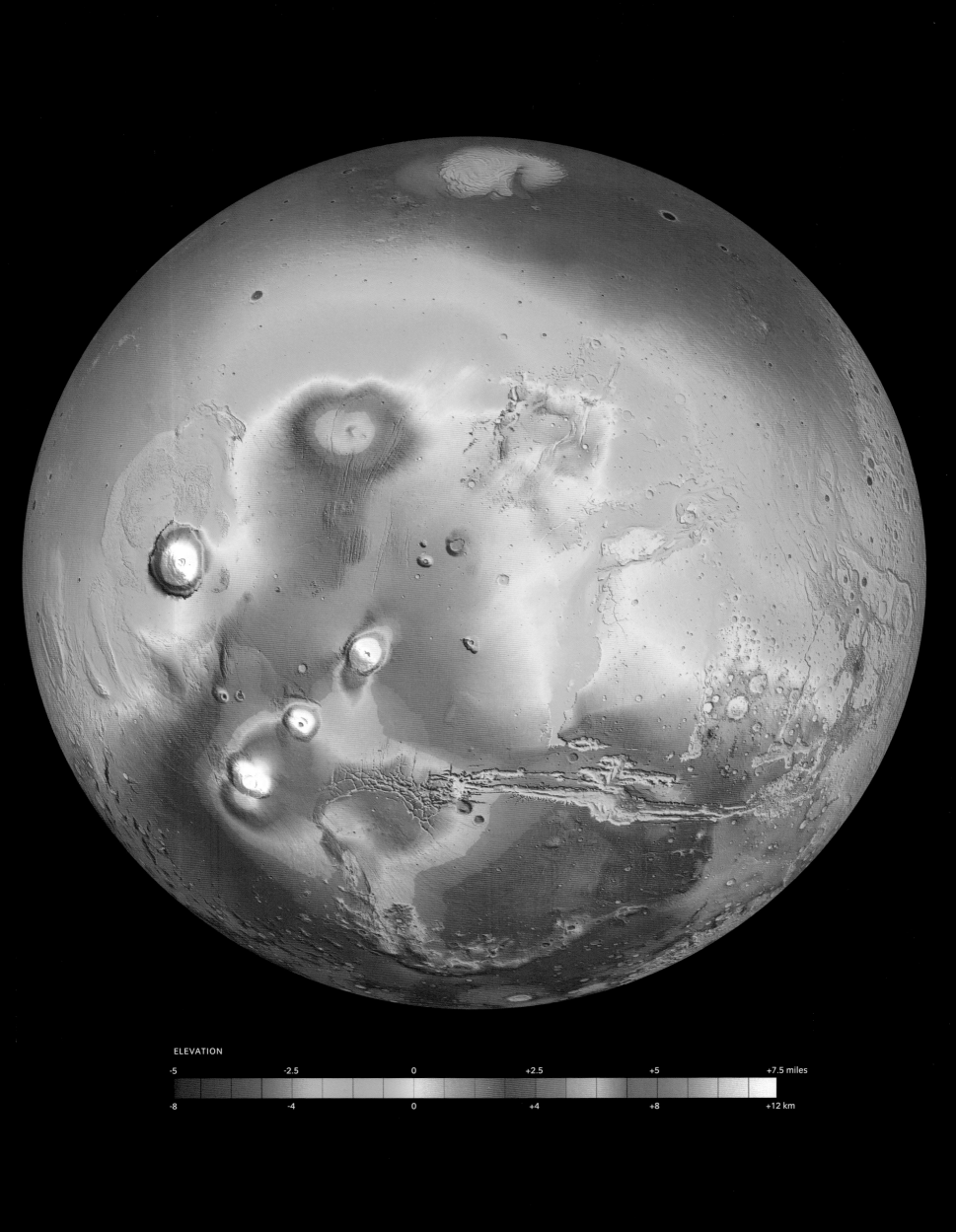

| -5 | | -2.5 | | 0 | | +2.5 | | +5 | | +7.5 miles |
| -8 | | -4 | | 0 | | +4 | | +8 | | +12 km |

The Martian Surface

ORBITING spacecraft yield global views, but the landers set themselves down on the planet's very surface, revealing a wealth of detail. The first successful lander, Viking 1, touched down in Chryse, a northern plain named for a far-off island rich in gold. The actual Chryse proved surprisingly rich in rocks—they covered about 8 percent of the surface. Viking 1 also saw dunes and the rims of distant craters. The dunes must be recent, because they align with the present wind direction.

Viking 2 set down over four thousand miles away, in northeastern Utopia Planitia, another northern plain. The scene there was more monotonous, strewn with rocks covering 14 percent of the surface. These were likely ejected from nearby impacts, especially the one that formed Mie, a crater 110 miles to the east that appears purple on the topographic map.

The Viking landers looked for life, and both found something that mimicked it. When they wetted the soil, it gave off oxygen, as a plant does. Trouble was, even when the Vikings heated the soil to deadly temperatures, it continued to emit oxygen, so whatever was "breathing" could hardly be alive. Instead, the soil probably bears peroxides, such as H_2O_2, which when wet shed oxygen. Despite the chemical similarity to H_2O, peroxides are bad news for life, because they destroy organic compounds. So do the Sun's ultraviolet rays, which pierce the thin, nearly ozoneless atmosphere. Both the peroxides and the ultraviolet radiation explain why Viking found no organic molecules, even though meteorites must carry such material to the planet's surface.

The two spacecraft measured the soil's elemental composition. For perspective, let's first examine the Earth. Elementally, the terrestrial surface is rather unimaginative. Although nature created ninety-two elements from hydrogen to uranium, just two ele-

ments weigh down nearly three fourths of the Earth's crust. Surprisingly, the most common element in the rocks beneath our feet is the one we breathe: oxygen makes up 47 percent of the terrestrial crust's weight.

The next most common element in the terrestrial crust, accounting for 28 percent, is silicon. Science fiction writers have long imagined worlds bearing life based not on carbon but on silicon. That's because silicon resides in the same column of the periodic table as carbon and therefore shares some of its chemical properties. Against this idea, however, stands the Earth itself. Even though the Earth's surface has far more silicon than carbon, no silicon-based life developed. Furthermore, silicon joins with oxygen to form rocks—in particular, silicate minerals such as feldspar, quartz, and mica—which lack the flexibility that life seems to require.

Aside from oxygen and silicon, only six other elements contribute more than 1 percent to the Earth's crust. The first, the lightweight metal aluminum, makes up 8 percent of the terrestrial crust. Next is the element that colors blood, and the Martian surface, red—iron, which makes up 5 percent of the Earth's crust. Another element essential to human life, calcium, makes up 4 percent. Rounding out the list are sodium, potassium, and magnesium, each contributing 2 to 3 percent. Thus, out of ninety-two elements to choose from, the Earth's crust has picked a mere eight—oxygen, silicon, aluminum, iron, calcium, sodium, potassium, and magnesium—to make up over 98 percent of its weight.

FACING PAGE: Viking 1 landed in Chryse Planitia and saw rusty rocks under a pink sky.

And Mars? Even though thousands of miles separated the two Viking landers, the soil's elemental composition was virtually the same at both. As on Earth, oxygen and silicon dominate, but the Martian surface has over twice the terrestrial percentage of iron. Indeed, after oxygen and silicon, iron is the most

ern plains may have been wet. On the other hand, the chlorine and sulfur may have originated as hydrochloric acid and sulfur dioxide spewed out by

Martian Meteorites

SUCCESSFUL though these spacecraft were, none delivered Martian rocks to Earth. Yet terrestrial laboratories have scrutinized dozens of such rocks, because asteroid impacts have splattered Martian rocks into space. Some of these rocks then hit the Earth.

All the Martian meteorites are volcanic rocks. They first came to attention in the late 1970s, because of their peculiar youth. Most other meteorites, originating from asteroids, are as ancient as the solar system, 4.6 billion years old. The strange young meteorites were only about a billion years old, however, so they must have come from a world whose volcanoes erupted that recently. Asteroids couldn't be the source, since they're so small they quickly lost the heat of youth, nor could the Moon, whose volcanoes shut down billions of years ago. The only logical suspect was Mars. Furthermore, the strange meteorites were spiced with iron, like the Martian surface. The clincher came in 1982, when scientists found that some of the meteorites had trapped gases whose composition matched that of the Martian air.

Martian meteorites are often called SNC meteorites, after three of the many places where they fell: Shergotty, India, in 1865; El Nakhla, Egypt, in 1911; and Chassigny, France, in 1815. The second of these consisted of some forty stones, one of which reportedly killed a dog. (Scientists count one fall as only one meteorite, even if it splits into pieces.)

Because no one knows where on Mars they came from nor their geological context, the rocks are not as informative as those a spacecraft would return. Moreover, some sat on Earth for hundreds of thousands of years before their discovery, so terrestrial compounds may have contaminated them. Nevertheless, harvesting Martian meteorites is a lot cheaper than sending spacecraft to Mars. Every Martian mete-

orite represents an actual chunk from the red planet's surface; to touch one is to touch Mars itself.

Strangely, most Martian meteorites are young, whereas most of the Martian surface is old. All but one of the Martian meteorites are between 165 million and 1.3 billion years old, corresponding to the Amazonian period; one other, found in Antarctica and named ALH 84001, is 4.5 billion years old, dating back to the Early Noachian period. Perhaps old rocks don't travel well. Old material on Mars may be gravelly. When blasted into space, old rocks may disintegrate into debris that can't survive the fiery passage through the Earth's atmosphere.

All of the Martian meteorites are mafic volcanic rocks, similar to those at the two Viking sites. Their youthfulness reveals that Martian volcanoes have erupted as recently as 165 million years ago and probably still do today. Most of the meteorites probably come from Tharsis, because it is the largest volcanic province on Mars. As noted earlier, the Martian meteorites have also revealed that the red planet's iron core formed fast and that Mars was born hot.

The core once generated a magnetic field that shielded the early Martian atmosphere from the ravages of the solar wind, which tried to tear the air away. During the Noachian period, the Martian atmosphere may have been thick enough to warm and wet the ruddy earth below. As we see in AIR, the thin veil of gases surrounding Mars today still preserves clues to the planet's distant past.

FACING PAGE: Most Martian meteorites are young volcanic rocks, suggesting that the red planet's volcanoes still

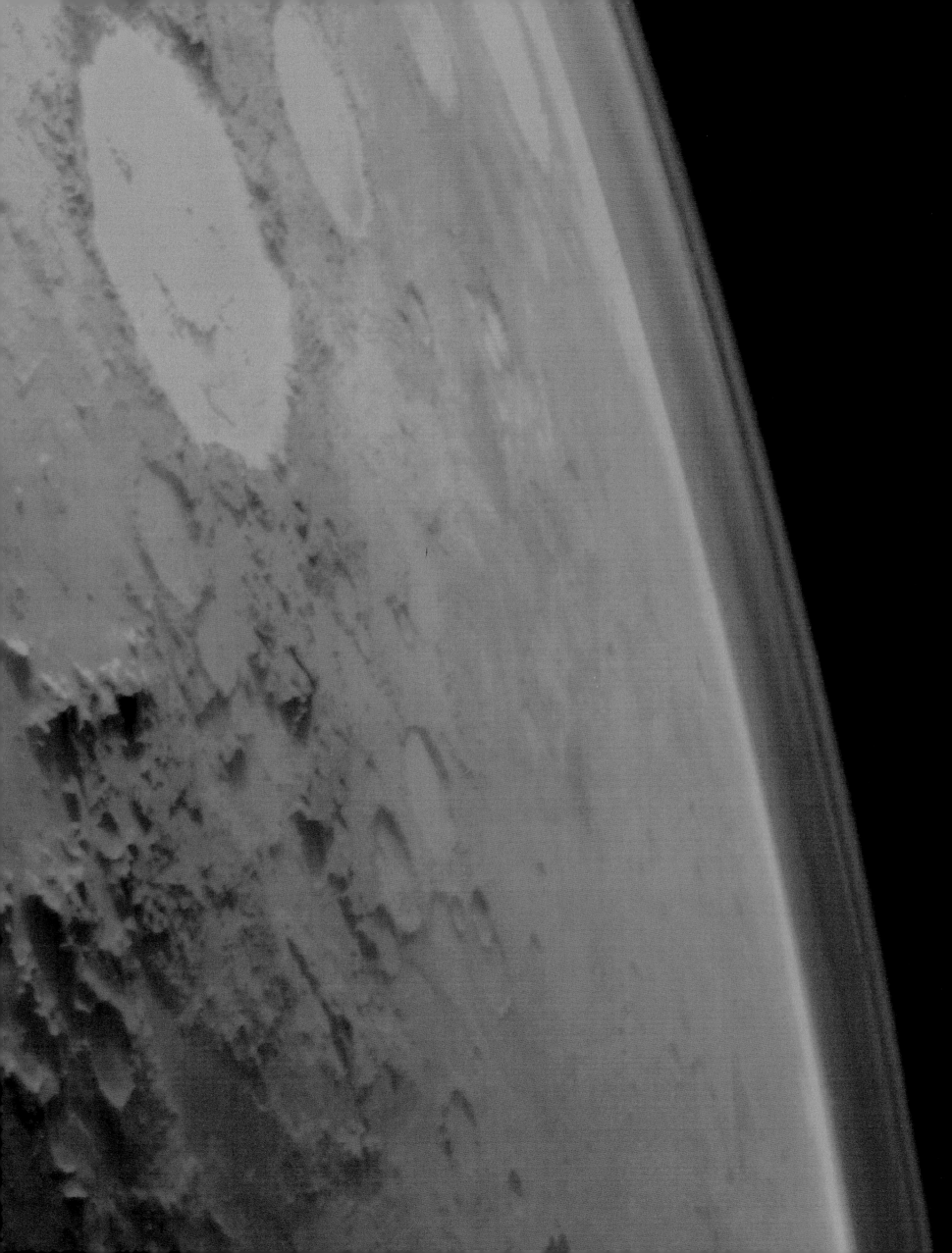

Air

WINDS RACE DOWN TALL MOUNTAINS, dust devils kick up desert debris, frost

forms on frigid polar soil: thin though Martian air is, barely warming the world it envelops, it never-

theless sports an arena of atmospheric activity. When Mars veers nearest the Sun, solar heat can whip

up dust storms that engulf the entire planet, while in winter up to a quarter of the atmosphere van-

ishes, freezing onto the dark polar ice cap. Furthermore, because Mars lacks a large moon to stabilize

its spin, its rotation axis oscillates, subjecting the planet to extremes of climate. Billions of years ago,

a thicker, wetter atmosphere—delivered by comets and asteroids, exhaled by volcanoes—may have

overseen the development of Martian life.

FACING PAGE: Dust hangs in the air above the Argyre impact basin. North is up.

In some ways Martian meteorology resembles the terrestrial. Both Earth and Mars are small worlds with air sufficiently thin that sunlight penetrates to the surface. As a result, the surface warms during the day and cools at night. Both planets rotate rapidly, and both have similar axial tilts and therefore similar seasons. On both planets carbon dioxide gas traps solar heat and warms the air.

But there are also key differences. Oceans cover most of Earth and none of Mars. Because oceans respond sluggishly to temperature changes, they moderate the terrestrial climate, as anyone who lives beside one can attest. Without oceans or a thick insulating atmosphere, the Martian surface suffers far greater temperature fluctuations. Furthermore, oceans ferry heat from the equator to the poles, so on Mars the thin air alone must do the job, and oceans trap dust and debris, abating terrestrial dust storms.

> In some ways Martian
> meteorology resembles
> 'the terrestrial

On Earth water vapor provides atmospheric power, because water switching from vapor to liquid releases energy that fuels storms; but on Mars dust plays this role, for airborne dust absorbs sunlight and heats the atmosphere. Mars has a more extreme topography, with higher highs and lower lows, modifying the Martian winds, and Mars but not Earth follows such an elliptical orbit that its varying distance from the Sun affects its seasons.

The Seasons of Mars

LIKE EARTH, Mars has seasons, and for the same reasons. Contrary to popular belief, the Earth does *not* have seasons because its distance from the Sun varies. Yet a poll of Harvard graduates found that many believed summer occurs when Earth is closest to the Sun and winter when farthest. More astonishing, at least one graduate student in Harvard's Department of *Astronomy* believed this, lending new weight to the saying, "You can always tell people from Harvard—but you can't tell them much."

The Earth's seasons actually arise because it does not stand perfectly upright. Instead, its axis is tilted 23.4 degrees from the vertical and points at Polaris, the North Star. Since Polaris is not directly "above" the Sun, during half of Earth's orbit the northern hemisphere is leaning toward the Sun. As a result, days in the north lengthen, nights shorten, and sunlight intensifies, causing the northern hemisphere to experience spring and summer while the southern hemisphere experiences fall and winter. Half a year later, when the north leans away from the Sun, the north has fall and winter and the south spring and summer. If instead the Earth's varying distance caused the seasons, both hemispheres would have the same seasons at the same time.

FACING PAGE: When the Sun sets on Mars, the sky finally turns blue—at least, around the Sun—because airborne dust scatters sunlight. On Earth the same phenomenon can cause a rare blue Moon. Higher up, Martian dust paints the sky pink.

The Earth's orbit is so circular that during a year its varying distance from the Sun alters the intensity of sunlight by only 7 percent. To the surprise of many, Earth is closest to the Sun in early January, when the northern hemisphere is having winter, and farthest in early July, during northern summer, further proof that the Sun's distance doesn't dictate the season. Logically, this variation in distance should intensify the southern seasons—since the Sun is closest during southern summer and farthest during southern winter—but the abundance of ocean water in the southern hemisphere so moderates the southern seasons that they are actually milder than their northern counterparts.

Nevertheless, the Earth's slightly elliptical orbit does produce one noticeable effect: people in the northern hemisphere see more of summer than of winter. That's because, as Kepler discovered by studying Mars, a planet moves slowest when farthest from the Sun. Since the Earth is farthest during northern summer, it moves slowest then, and northern summer lasts four and a half days longer than northern winter.

Even on Mars, whose orbit is far more elliptical, the cause of the seasons is the axial tilt, but over a Martian year the sunlight intensity varies by a whopping 45 percent. Mars is nearest the Sun just before the start of southern summer, so southern seasons are the more severe. This is probably why dust storms usually erupt in the south. Because of the elliptical orbit, the seasons vary in length much more than they do on Earth. The longest season, northern spring/southern fall, lasts 52 terrestrial days longer than the shortest season, northern fall/southern spring: 199 days versus 147.

Atmospheric Composition

MARTIAN AIR DIFFERS from terrestrial. Carbon dioxide makes up 95.32 percent of the red planet's atmosphere, over two thousand times the earthly percentage. Only Venus has a greater proportion. On Earth human beings and other animals produce this gas, but plants breathe it in and oceans dissolve it, so it makes up a mere 0.037 percent of the terrestrial atmosphere. To people carbon dioxide is a poison: levels of just a few percent kill.

At both terrestrial and Martian pressures carbon dioxide is never liquid. Instead, it goes directly from gas to ice and back again. On Mars it freezes around −195 degrees Fahrenheit. During southern winter so much carbon dioxide freezes onto the south polar cap that a quarter of the planet's atmosphere disappears. The Earth never suffers this problem, because it is warmer and its main gas, nitrogen, turns to liquid at −320 degrees Fahrenheit. As Mars approaches southern spring, the carbon dioxide ice in the south begins to vaporize, and some of the gas then freezes onto the northern ice cap, where autumn is advancing.

The second most abundant gas on Mars is the one that makes up nearly four fifths of terrestrial air: nitrogen, which accounts for 2.7 percent of the Martian atmosphere. Vital for terrestrial life, nitrogen occurs in protein, so its discovery in the Martian

FACING PAGE: One minute after sunset, the western sky is pink. Because dust extends high above Mars, the sky stays bright for up to two hours after sunset.

atmosphere augured well for the prospects of past life on the red planet. However, most organisms can't use molecular nitrogen (N_2), the dominant form in terrestrial and Martian air, because it's so stable. Instead, molecular nitrogen must be converted to another form, such as ammonia (NH_3) or nitrates. Lightning does this, as do bacteria using the enzyme nitrogenase. Unfortunately, no one yet knows whether the Martian surface has nitrates or other nitrogen compounds.

Surprisingly, Mars has almost the same percentages of carbon dioxide and nitrogen in its air as Venus, although the Venusian atmosphere is thousands of times thicker. Even Earth has a similar proportion of these two gases, if one counts the carbon dioxide locked in carbonate rocks like limestone. This similarity presumably reflects the similarity in the planetesimals that formed these three neighboring planets.

After carbon dioxide and nitrogen, the next most common gas in Martian air is argon, an inert element that makes up 1.6 percent (versus 0.93 percent of Earth's). At one time scientists thought argon might be a major atmospheric ingredient. That's in part because of erroneous measurements by a Soviet spacecraft, but also because radioactive potassium-40 in rocks can decay into argon-40. Since argon is inert, it does not react with the surrounding rocks. Instead, it creeps upward, to the surface, and then into the atmosphere. The rate of this so-called outgassing, however, is much lower on Mars than on Earth, because for most of its life Mars lacked the continental drift, quake activity, and vigorous volcanic eruptions that shake gas out of the ground. Scientists calculate that the pittance of atmospheric argon implies an outgassing rate only 1/16 of Earth's.

FACING PAGE: An hour and forty minutes before sunrise, pink clouds ten miles high blanket the Martian sky. They consist of water-ice crystals adhering to reddish dust.

Although oxygen is the most common element in Martian soil, it amounts to a mere 0.13 percent of Martian air

The gas human beings breathe is even rarer. Although oxygen is the most common element in Martian soil, it amounts to a mere 0.13 percent of Martian air. Oxygen has only itself to blame. It's so reactive that it quickly combines with the rocks and removes itself from the air. Earth's air has oxygen only because the planet below has life. If all terrestrial life perished, atmospheric oxygen would vanish just 4 million years later.

Because Mars has little oxygen in its air, it also has little ozone (O_3), which forms when oxygen atoms (O) join oxygen molecules (O_2). Ozone in the terrestrial stratosphere blocks the Sun's deadly ultraviolet light. According to a 2000 study, the Martian surface gets hit by ten times more of the deadliest forms of ultraviolet radiation than the Earth's surface. Modern Mars is caught in a vicious cycle: no life means no atmospheric oxygen, no atmospheric oxygen means no ozone, and no ozone means no life, at least on the surface. The little Martian ozone occurs when least needed, in winter, under weak sunlight. That's because cold air has less water vapor, which helps destroy ozone.

Because Mars has almost no ozone, it also has no stratosphere. Earth has a stratosphere because ozone absorbing the Sun's ultraviolet radiation heats this region and creates a temperature inversion. Normally, the higher you go, the colder the air gets; but in the stratosphere, climbing higher leads to hotter air, not colder. Such a temperature inversion causes extreme stability: if you lift a parcel of air in the stratosphere, it finds itself colder and thus heavier than its surroundings, so it sinks back to its original level. The air below the stratosphere, the so-called troposphere, is just the opposite and therefore triggers lively weather. On stratosphereless Mars the troposphere extends right up to the mesosphere, the middle atmosphere. Above that, on both planets, the temperature rises with height, because the Sun's extreme ultraviolet radiation heats the few atoms in the exosphere, the upper atmosphere.

Until recently, the low oxygen abundance in the Martian atmosphere seemed peculiar, because sunlight should split some of the abundant carbon dioxide (CO_2) into carbon monoxide (CO) and oxygen (O). In 2001, however, a NASA spacecraft orbiting Earth, the Far Ultraviolet Spectroscopic Explorer (FUSE), resolved the puzzle by detecting molecular hydrogen (H_2). There wasn't much—only about 15 parts per million—because this light gas readily escapes into space. Still, the chemical reactions that produce molecular hydrogen involve compounds containing one hydrogen atom, such as H, OH, and HO_2, which also convert oxygen and carbon monoxide back into carbon dioxide, explaining why Martian air can have so much CO_2 but so little CO and O.

In 1993 another Earth-orbiting spacecraft, NASA's Extreme Ultraviolet Explorer (EUVE), had discovered another light gas. Helium, like hydrogen, quickly escapes the planet, and its abundance is only about 4 parts per million, or 0.0004 percent. A 2000 study by Vladimir Krasnopolsky of the Catholic University of America, who led the team that discovered the helium, estimated that 70 percent of the gas comes not from Mars but from the Sun. The Sun is mostly hydrogen and helium, and it shoots out a stream of charged particles called the solar wind, which mixes with the upper Martian atmosphere. The remaining 30 percent of Martian helium, Krasnopolsky said, emerges from the ground below. The mantle and crust contain radioactive thorium and uranium, which decay into lead and give off helium. Like argon, helium is an inert gas that does not react with the surrounding rocks but instead creeps upward and eventually outgasses into the atmosphere. From the scarcity of helium in the red planet's air, Krasnopolsky estimated that the present Martian outgassing rate is only 1/15 that of the Earth, consistent with the low outgassing rate derived from the argon abundance.

The Martian atmosphere has other noble gases, too. The Viking landers detected traces of neon, krypton, and xenon. Since noble gases don't get locked into molecules, they and their isotope ratios give glimpses of the ancient Martian atmosphere and thus lend clues to the ancient Martian climate, a property we exploit at this chapter's end.

Because Mars has almost no ozone, it also has no stratosphere

FACING PAGE: Forty minutes before sunrise, water-ice crystals in these wispy eastern clouds forward scatter sunlight, coloring themselves blue.

permit visible sunlight to warm the ground, but when the ground radiates that warmth, at infrared wavelengths, the gases try to block its escape into space. The greenhouse effect warms the world to a balmy 60 degrees Fahrenheit.

Mars isn't so lucky. Although carbon dioxide dominates the atmosphere—a column of Martian air contains about forty times more carbon dioxide than

The greenhouse effect raises the Martian temperature by only about 10 degrees Fahrenheit

does a similar column on Earth—this gas isn't as potent as water vapor, which the Martian atmosphere largely lacks. As a result, the greenhouse effect raises the Martian temperature by only about 10 degrees Fahrenheit. The feeble greenhouse effect and the great distance from the Sun conspire to curse Mars with a mean temperature of −67 degrees Fahrenheit. Thus, on the Fahrenheit scale, Martian temperatures are as negative as terrestrial temperatures are positive.

Both Viking and Pathfinder felt the Martian freeze firsthand. They were coldest just before dawn,

slowly. At the Pathfinder site, where it was summer, the temperature each day sank to around −100 degrees Fahrenheit and peaked at around +10 degrees Fahrenheit—a daily temperature range of over 100 degrees F., a consequence of the thin air that barely insulates the surface.

Over the entire planet, the temperature gets coldest at the pole experiencing winter darkness. There the temperature drops to the freezing point of carbon dioxide, −195 degrees F. It can get no colder than this because when carbon dioxide gas freezes, it actually releases heat. Thus, if the ground tried to cool further, additional carbon dioxide would freeze out of the air and thereby warm the ground, maintaining a temperature of −195 degrees F. The planet gets warmest during late southern spring in the part of the southern hemisphere where the Sun is overhead. This is when Mars comes closest to the Sun. There the noon temperature at the planet's surface flirts with +80 degrees F.

Cold though it was, the Pathfinder spacecraft discovered that during the day the air just a few feet higher was several degrees colder. That's because sunlight warms the ground, but the thin air fails to carry the heat far upward. If you've ever trotted barefoot across sunlit asphalt, you've felt the same phenomenon. Because the sunlight striking the Martian surface controls the temperature, a Martian mountaintop may be just as warm as the valley below—a striking contrast with Earth.

The Viking and Pathfinder spacecraft also measured the atmospheric pressure. On Earth this is around 1 bar, or 1,000 millibars. The Martian average is only 6 millibars, 1/170 of the terrestrial value. Because of the weak gravity, however, Martian air doesn't press down hard, so to exert a given pressure Mars needs more air above the surface. As a result, the average mass in a vertical column of Martian air is not 1/170 of the terrestrial value but about 1/70. Air at the Martian surface is as rarefied as the terrestrial stratosphere.

The atmospheric pressure varies with both location and season. Air, like water, seeks the lowest level, so Martian air is thickest in low-lying regions such as the northern plains and the southern impact basins Hellas and Argyre; it is thinnest atop the tallest mountains. Because of the red planet's great topographic range, discussed in EARTH, the atmospheric pressure in the deepest depths is over ten times that atop the mightiest mountains. The Viking and Pathfinder spacecraft all landed in the northern lowlands, so they recorded pressures higher than the planetary average.

Because a quarter of the atmosphere vanishes onto the south polar cap during southern winter, the atmospheric pressure also varies with season. This does not happen on Earth. Since southern Martian seasons are more extreme than their northern counterparts, the south polar cap's size largely determines the atmospheric pressure all over Mars. During southern summer solstice, when sunlight shines strongest on the south pole, carbon dioxide ice there vaporizes, adding gas to the atmosphere. That's when the atmospheric pressure peaks. During southern winter, when the south polar cap plunges into darkness and carbon dioxide freezes onto it, the atmospheric pressure plummets, reaching its annual minimum.

> Martian air is thickest in low-lying regions such as the northern plains and the southern impact basins Hellas and Argyre

Although the Viking spacecraft landed in the northern hemisphere, both witnessed this south-pole-driven seasonal variation. During a Martian year the atmospheric pressure at the Viking 1 site went from about 6.8 millibars at minimum to 9.0 millibars at maximum—a variation of 30 percent. On Earth the atmospheric pressure rarely varies by more than 5 percent. Because Viking 2 landed in the Utopia impact basin, it recorded higher atmospheric pressures, from 7.4 millibars at minimum to 10.2 millibars at maximum.

Thin though the Martian air is, it nevertheless burns up small meteoroids that try to strike the surface. This may be why Mars has so many more rocks than the Moon—micrometeorites haven't pulverized them. As a result, when Martian colonists look skyward at night, they should enjoy meteors and even meteor showers, the latter when the red planet plows through debris shed by a comet. Larger objects, however, survive their passage through the atmosphere. A 2000 study estimated that the average square mile of

On Earth most water vapor comes from the warm subtropics, whereas on Mars it comes from the frigid polar caps

Martian territory bears between 1,300 and 1.3 million meteorites heavier than a third of an ounce. Some are probably from Earth. Imagine if astronauts found fossils in a Martian rock that actually came from Earth!

Martian air is dry, yet the relative humidity is high, so clouds form. Although this sounds like a contradiction, it merely reflects the extreme cold. Frigid air can't hold much water vapor, so even a trace boosts the relative humidity to 100 percent and creates clouds. This is why clouds on Earth often form when air rises. As the rising air reaches lower pressures, it expands and cools, and what had been a low relative humidity becomes higher. Conversely, when air sinks, clouds tend to dissipate. Low-pressure centers have rising air, so on Earth they bring clouds and rain, whereas high-pressure centers have sinking air, so they bring fair skies.

The Martian atmosphere is so dry that if the water vapor were wrung out of it and spread all over the surface, the resulting "ocean" would be only 10 microns thick—definitely not an ocean to scuba dive in. In contrast, the wet terrestrial atmosphere would yield a global ocean one inch deep, or thousands of times thicker. The two planets get their water vapor from opposite places. On Earth most water vapor comes from the warm subtropics, whereas on Mars it comes from the frigid polar caps.

Clouds often form in the Martian mountains, such as the mighty volcanoes of Tharsis and Elysium. During spring and summer mornings and afternoons, moist air flows uphill over these mountains, cooling until the relative humidity reaches 100 percent and clouds form. Clouds also form over the polar caps when they are expanding, during autumn and winter. Most Martian clouds are made of water ice; some of the polar clouds, though, may be carbon dioxide ice. The Pathfinder spacecraft saw clouds in the early morning, but these vanished as the temperature rose.

Like Earth, Mars has cyclones—low-pressure storms—but they don't produce rain or snow. They do affect the winds, however. These dry storms passed over the Viking landers. In 1999 the Hubble Space Telescope captured a giant cyclonic storm that looked like a terrestrial hurricane spanning over 1,000 miles; it even had an eye 200 miles across. The storm erupted near the edge of the north polar cap, probably because of the contrast between the cold ice and the relatively warm surrounding soil. Like low-pressure storm centers in Earth's northern hemisphere, the storm spun counterclockwise.

FACING PAGE: A cyclonic storm whose eye spans 200 miles swirls to the lower left of the north polar cap. North is up.

Martian Winds

OUTSIDE OF STORMS, Martian winds tend to be light, especially at night. At both Viking sites the nighttime wind speed was about 4 miles per hour. During the day, the winds increased to around 15 miles per hour. Nighttime winds tend to be downslope, from high elevations to low, and daytime winds upslope. For example, at the Pathfinder spacecraft, which landed near the mouth of Ares Vallis, winds blew down the valley at night, then up the same valley during the afternoon.

Even from afar, the wind makes its mark; many craters bear wind streaks. Some streaks are bright, caused by bright dust deposited downwind from the crater. Other streaks are dark, either because the wind has deposited dark material or it has eroded light surface material and exposed darker rock below. Some crater floors have dark deposits, too—probably material that the wind can barely move and that can't climb over the crater rim, so it gets stuck.

The wind streaks indicate the wind direction

The wind streaks are useful because they indicate the wind direction. They point in the directions expected to prevail during southern summer and early fall, when the great dust storms are ending. At northernmost latitudes—poleward of 40 degrees north—the wind streaks indicate west winds; from 30 degrees north to the equator, northeasterly; from the equator to 20 degrees south, northerly; from 20 degrees south to 30 degrees south, northwesterly; and poleward of 30 degrees south, easterly or southeasterly.

Mars exhibits other signs of the wind. Yardangs are streamlined ridges that on Earth form when strong winds erode fine-grained sedimentary rock. They occur in all great terrestrial deserts, but geologists largely ignored them—until spacecraft saw them on Mars. Like their earthly twins, Martian yardangs align with the prevailing wind. They are most common in the southern part of Amazonis Planitia, southwest of the giant volcano Olympus Mons, whose lofty peak must affect the Martian winds. Some Martian yardangs are thirty miles long.

The winds of Mars also sculpt enormous sand dunes. Most of Earth's sand dunes populate deserts at low to middle latitudes, but the greatest Martian sand dunes surround the north pole, covering an area as large as Texas. According to the Mars Global Surveyor's laser altimeter, the average dune here is eighty feet high, enough to bury a six-story building. From one crest to the next is a mile and a half. Dunes also appear in the south of Mars, but they are smaller, usually trapped inside craters. Unlike their earthly counterparts, most Martian dunes are dark. The Earth's dunes are usually light-colored because they're made of quartz sand; its few dark dunes—as in Hawaii—are made of broken-down basalt. Dark dunes don't last long on Earth, because the minerals in basalt weather fast. Since Mars experiences no rain and thus little weathering, its dark dunes can survive.

FACING PAGE: Easterly and northeasterly winds deposit bright streaks of dust downwind from small craters in Syrtis Major, the most prominent dark region of Mars. North is up.

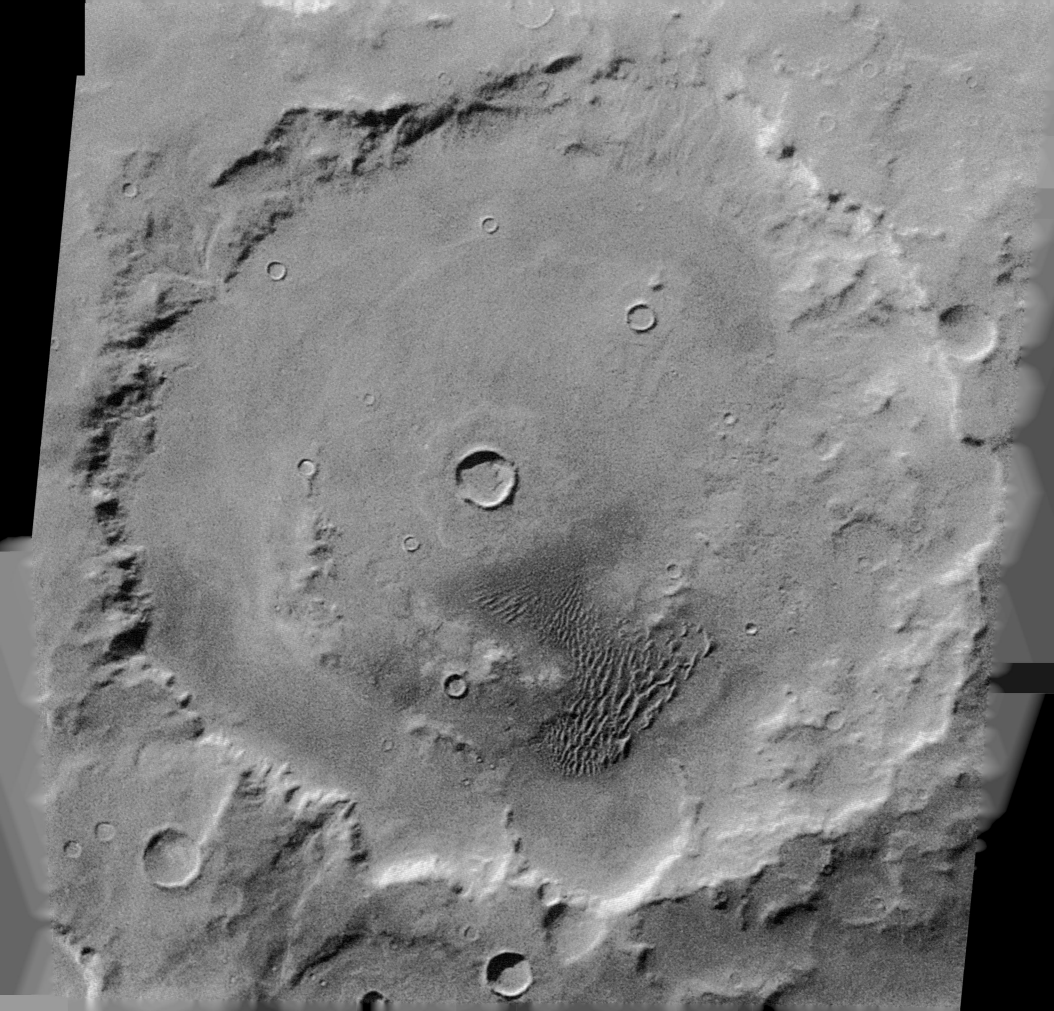

for example, raised the temperature of the Martian atmosphere by as much as 80 degrees Fahrenheit. And climatically, dust darkens the polar caps and promotes vaporization, since dark ice vanishes faster than bright.

The greatest dust storms usually begin when Mars nears the Sun, in late southern spring or early summer, and the southern hemisphere is the usual launching point. For example, the 1971 dust storm started in Noachis Terra and the 2001 dust storm in the giant southern impact basin Hellas. When Mars is closest to the Sun, intense sunlight whips up winds from gas to liquid—when a cloud forms out of thin air—it gives off heat. You feel the reverse of this process when water evaporates and cools your skin.

When a dust storm ends, the dust, which is bright, settles all over Mars, reducing the contrast between the planet's bright and dark areas. The contrast returns when winds sweep the terrain clean. Long ago observers witnessed these changes but optimistically attributed them to changing vegetation. By mixing material around the planet, dust storms may explain why, as noted in EARTH, the two Viking landers found that Martian soil separated by thousands of miles had the same composition.

In addition to dust storms, Mars has dust devils—swirling columns of dust that appear in terrestrial deserts on hot days and somewhat resemble tornadoes. On Mars dust devils even passed over the Viking and Pathfinder spacecraft.

FACING PAGE: Dark dunes blemish southern sections of Kaiser Crater, one of the many craters in Noachis Terra. North is up.

FOLLOWING PAGES: An enormous dust storm swept over Mars in 2001. North is up.

Page 86: June 26, 2001: Now you see Mars.

Page 87: September 4, 2001: Now you don't.

Climate Cycles

MARS GOES through climate cycles. So does Earth. The most dramatic are the ice ages, when glaciers grind out of the north and bury vast tracts of land beneath ice up to a mile thick. The ice ages are thought to result when the other planets cause small changes in the Earth's axial direction, axial tilt, and orbital shape that conspire to cool northern summers, so ice accumulates and glaciers stampede south. Fortunately for terrestrial inhabitants, the Moon's gravity stabilizes the Earth's axis and prevents more devastating changes.

Mars lacks a large moon, so its climate cycles are far more severe. There are three of them, whose lengths scientists can calculate by considering the gravitational pull of the other planets on Mars. The shortest cycle is the precession cycle, which on Earth lasts 23,000 years and on Mars 51,000 years. Precession causes a planet to wobble, altering where its axis points—in particular, which hemisphere leans sunward when the planet nears the Sun. Right now, because Mars is nearest the Sun during late southern spring, southern seasons are more extreme. But in 25,500 years—half the precession cycle—this will be reversed: Mars will be closest to the Sun during late *northern* spring, so northern seasons will be the more extreme.

The second cycle, probably the most crucial, alters the planet's axial tilt. Earth's present tilt is 23.4 degrees, and the red planet's is similar, 25.2 degrees. But the axial tilt oscillates. When the Martian axial tilt is low, sunlight barely reaches the poles and little ice vaporizes. Atmospheric circulation weakens, and the air may thin, causing dust storms to cease. At such times, carbon dioxide may no longer dominate the atmosphere, because so much freezes onto the polar regions. On the other hand, when the axial tilt is great-

est, sunlight bathes the poles, invigorating atmospheric circulation and presumably launching ferocious dust storms. In addition, if the polar sunlight vaporizes more ice, the thicker air promotes dust storms, too, because thick air more readily lifts dust.

This axial tilt cycle lasts 41,000 years on Earth, 120,000 years on Mars. Thanks to the lunar anchor, however, the Earth's axial tilt varies by less than 2 degrees, going from about 22.5 to 24.4 degrees. However, nearly moonless Mars suffers an enormous variation. By chance, its present axial tilt of 25.2 degrees is close to its average. But during one cycle, the axial tilt can oscillate by over 10 degrees. During just the last several million years, the Martian axial tilt has varied all the way from 13 degrees to 42 degrees.

That's bad enough, but in times past Mars may have experienced axial tilts as low as 0 degrees—no tilt at all—and as high as 60 degrees. On Earth and present Mars more sunlight illuminates the equator than the poles. That's why the equator is warmer. But as the axial tilt increases, more sunlight illuminates the poles. When the axial tilt climbs to 54 degrees, the poles get as warm as the equator, and when the axial tilt exceeds 54 degrees, as for Uranus and Pluto, the poles actually become warmer than the equator. If the Earth were so tilted, ice would blanket Africa and heat would sear Antarctica.

FACING PAGE: The Moon guards Earth's climate. Mars isn't so lucky.

The third and final climate cycle alters the orbital eccentricity, the shape of the planet's orbit around the Sun. At the moment the Martian orbit is quite elliptical, with an eccentricity of 9.3 percent. But the eccentricity varies greatly, with a long period of 2 million years and a short period of about 100,000 years. During the 2-million-year period, Mars goes from zero orbital eccentricity—that is, a circular orbit—to an orbital eccentricity of 12 percent. When Mars follows a perfect circle, the intensity of sunlight doesn't change over a Martian year; perhaps great dust storms cease, since Mars never gets hot enough to stimulate them. When Mars pursues its most elliptical of orbits, sunlight shines 60 percent more intensely when the planet is closest than when farthest. At such times, even a casual observer on Mars would notice that the Sun looks fainter when farthest. A high orbital eccentricity exacerbates the extremity of the seasons and presumably the severity of dust storms: whichever hemisphere has summer when Mars comes closest to the Sun suffers scorching summers and frigid winters.

Because of their severity, the climate cycles may have helped plunder the planet's ancient atmosphere, dry up its water, and kill its life. No other inner planet suffers like Mars. Mercury and Venus stand almost perfectly upright, because they are so close to the Sun that its gravity stabilizes them, and the Earth is lucky that a large satellite protects it—at least, for now. The Earth faces future trouble because the lunar influence is fading. That's because the Moon is receding. As a result, according to calculations by William Ward of the Southwest Research Institute in Boulder, Colorado, the terrestrial axial tilt will increase, warming the poles and cooling the equator. Over the next 1.5 billion years, Ward says, the Earth's mean axial tilt will inch up to 26 degrees, then jump to 34 degrees. Half a billion years later, it will swell to 51 degrees, with excursions as high as 60 degrees. To

prevent the resulting climate changes, our descendants may want to haul the Moon back in.

Conveniently, Mars may have preserved a record of its recent climate cycles for scientists to study. On Earth climatic fluctuations alter the growth of trees, which preserve these fluctuations in tree rings. Although Mars has no trees, the north and south polar regions bear alternating bright and dark layers that resemble scaled-up tree rings. Each pair is 30 to 160 feet thick, and some run for hundreds of miles. They spiral outward from the poles, counterclockwise in the north, clockwise in the south, down to about 80 degrees latitude in the north and 70 degrees in the south. At each pole these polar layered deposits cover an area three fourths as large as Alaska. They probably consist of ice and dust in differing amounts—the brighter layers have more ice, the darker ones more dust. They are very young—Late Amazonian in age—and presumably were laid down over the past several million years. Unfortunately, since no spacecraft has sampled them, the exact ages of the sequences are unknown, as is the particular climate cycle—precession, axial tilt, or eccentricity—they reflect. Broadly, though, when climatic conditions promote dust storms, such as during times of high axial tilt, the ice probably becomes dirty, forming a dark layer; when dust storms cease, as during times of low axial tilt or low orbital eccentricity, then the ice stays clean, and a bright layer results.

FACING PAGE: Polar layered deposits—the alternating brown and tan layers seen here just right of the white ice— probably record recent changes in Martian climate. North is to the left.

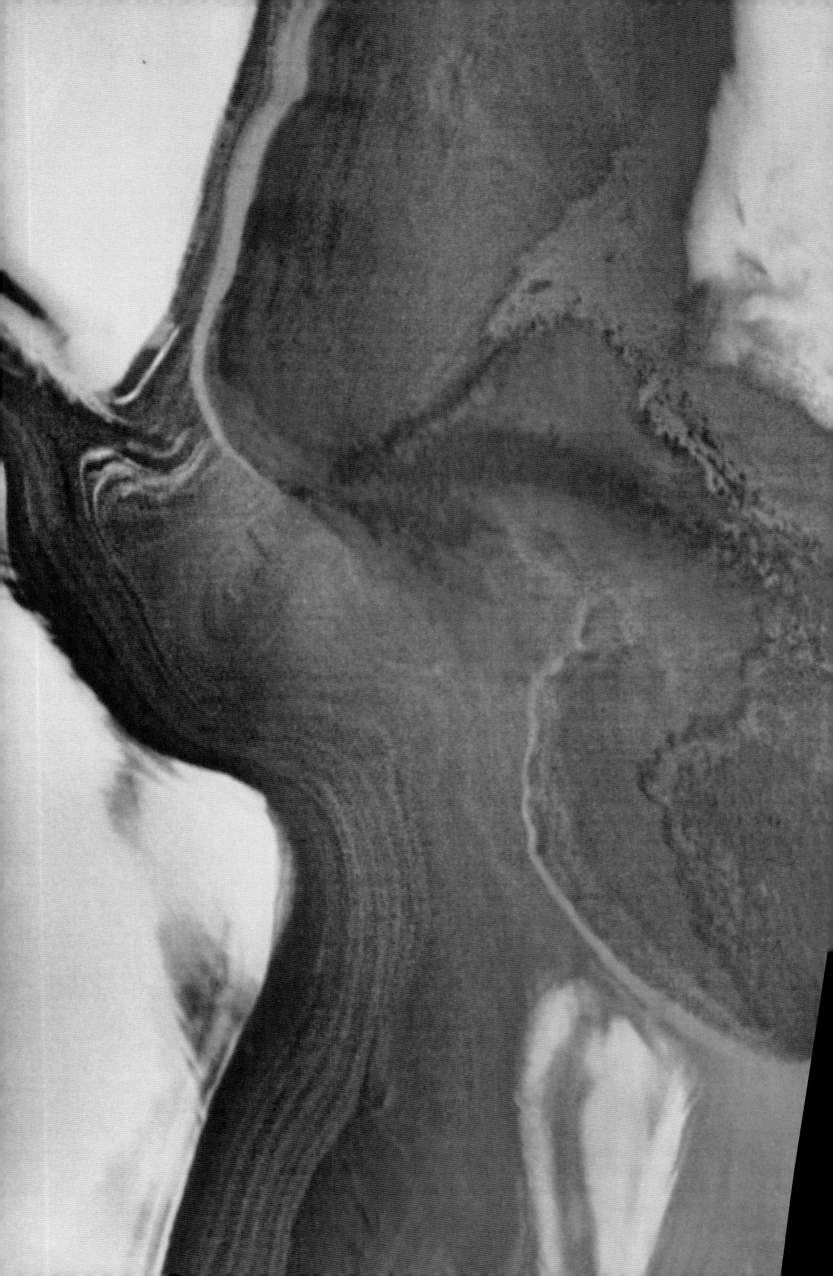

The Vanishing Act of Martian Air

IF MARTIAN VOLCANOES ever erupted, if Martian rivers ever flowed, if a Martian ocean ever lapped the northern plains, if Mars ever gave birth to life, it surely had more air than it does today. Volcanoes cough up air, which rivers, lakes, and oceans require; otherwise, Martian air grows so cold and thin that water freezes and then vaporizes. Indeed, through the greenhouse effect, an atmosphere boosts the surface temperature—especially vital at the red planet's distance from the Sun—helping water to flow. During the Noachian period, rivers *did* flow on Mars, as their dried-up remains in the cratered highlands testify, and worn-down craters indicate an erosion rate a thousand times today's. Thus, ancient Mars had much more than the gossamer of gases surrounding it now.

Unfortunately, Mars lost most of its atmosphere. No other planet is known to have suffered such a catastrophic transition. The atmosphere of infant Mars may have been a dozen or so times thicker than the present Earth's and composed mostly of hydrogen, according to the University of Minnesota's Robert Pepin. This hydrogen didn't come from the primordial disk whirling around the newborn Sun—Mars was too small to capture much of the lightweight element—but instead from comets smashing into the planet. The comets carried water (H_2O) that split into molecular hydrogen, which filled the air, and oxygen, which rusted the rocks. However, other scientists, such as the University of Hawaii's Tobias Owen, doubt that Mars ever had much hydrogen in its air.

Whatever the case, any hydrogen-rich atmosphere didn't last long. In fact, it was gone within 100 million years of the red planet's birth—the early years of the Early Noachian period. That's because molecular hydrogen absorbed extreme ultraviolet radiation from the Sun and got heated. This heat so increased hydro-gen's speed that it escaped into space. Furthermore, if this hydrogen-dominated atmosphere was as thick as Pepin suspects, then the escaping hydrogen would have dragged heavier gases with it. According to Pepin, such an escape can explain the abundances of xenon isotopes in the Martian air. Taking its name from the Greek word for "strange," xenon is an inert gas, so it stays free of other elements and helps trace the history of the atmosphere. Furthermore, with an atomic number of 54, xenon is more than heavy enough for even a small planet like Mars to retain. Yet when the Viking landers measured the abundances of the xenon isotopes, they found that the lighter isotopes are rarer in Martian air than terrestrial air. This suggests that hydrogen's escape from Mars preferentially dragged the light xenon isotopes into space.

Still, Mars had an atmosphere in reserve, beneath its surface, locked in the rocks of its mantle and crust. During the Noachian period, as volcanoes erupted, their lava spewed gases such as carbon dioxide and nitrogen into the atmosphere. The lava was wet, so water vapor entered the air, too. If Mars ever had continental drift, the churning of the crust also helped inject gases from the ground into the air. As a result, the early Martian atmosphere probably resembled early Earth's, consisting mostly of carbon dioxide, nitrogen, and water vapor. The carbon dioxide and water vapor were powerful greenhouse gases that trapped solar warmth and boosted the surface temperature. In the Earth's case, the greenhouse effect kept the planet from freezing, as it should have, because the young Sun was 30 percent fainter than it is today. How warm Mars got, no one knows.

But trouble lay ahead. Even as volcanoes were erupting, asteroids and comets were pummeling Mars. Small impacts delivered additional gases to the planet, but in a 1989 paper H. Jay Melosh and Ann Vickery of the University of Arizona proposed that giant impacts—such as those which carved the Hellas and Argyre basins—blasted gases into space, thereby stripping the planet of atmosphere. Because Mars is so small, the loss may have been as bad as 99 percent.

The atmospheres of Venus and Earth were safer from such attacks, because these planets were larger and their greater gravity better held on to their air, perhaps explaining why they have much thicker atmospheres than Mars today. Nevertheless, the Earth may have lost much of its original atmosphere when the Moon formed. Scientists think that the Moon arose when a Mars-sized impactor smashed into Earth and splattered debris from its mantle into orbit. This impact may have so heated the Earth's surface that it drove off most of the air—possibly a good thing, because otherwise the Earth's atmosphere might have remained as stifling as the one that suffocates Venus. On the other hand, the Earth's air may have thinned even without the Moon's birth, as oceans and rainfall removed carbon dioxide from the atmosphere and eroded rocks, binding the carbon into carbonate rocks such as limestone.

Small Mars, though, needed all the air it could get. The ancient Martian atmosphere may have oscillated between thick and thin, as the frequent small impacts delivered gases to the planet and then the rare giant impacts blew them off again.

Even as the Martian atmosphere was under attack from above, it was facing problems from below. When the Martian interior cooled, the volcanism in Tharsis and elsewhere that had been supplying gases presumably dwindled, reducing the input of new air. Also, if the planet had continental drift, that came to

Small Mars needed all the air it could get

an end, too. For another thing, as described in EARTH, ancient Mars had a magnetic field: the old regions of Terra Cimmeria and Terra Sirenum are magnetized. The magnetic field protected the Martian atmosphere from the solar wind, whose particles try to tear away the atmosphere. But the magnetic field lasted just a few hundred million years. When it vanished, the Martian air became easy prey for the solar wind. Solar-wind stripping probably peaked late in the Noachian period and early in the Hesperian period, just the time the climate was deteriorating. That may be no coincidence, because when the atmosphere began to degenerate, so did the climate it nourished through the greenhouse effect.

Furthermore, a thinner atmosphere had a harder time ferrying heat from the equator to the poles, so the poles cooled. In 1994, Robert Haberle of NASA's Ames Research Center, near San Francisco, and his colleagues proposed that this polar cooling led to another calamity. When the poles cooled to the freezing point of carbon dioxide, the gas began to freeze and form ice caps. Thus, the air thinned more, further impeding the equator-to-pole heat flow and further damping the greenhouse effect; the polar regions therefore cooled further and stole yet more air from the atmosphere, which only made things worse.

Most of these problems—the dwindling of volcanism, the cessation of any continental drift, the decay of the magnetic field, the freezing of the atmos-

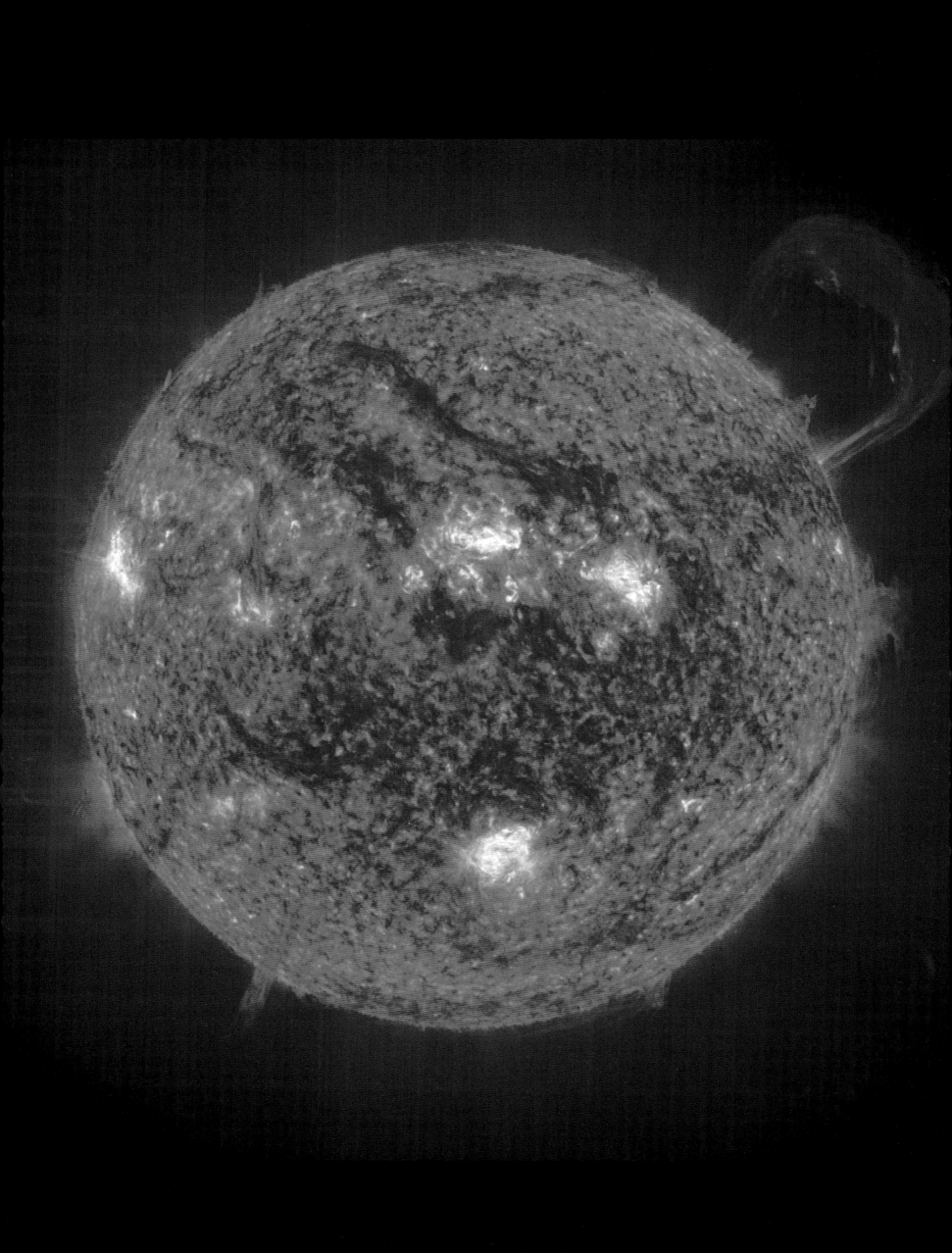

phere—had the same root cause: the small size of Mars, causing its interior to cool far faster than Earth's. Furthermore, it was the planet's small size that left the atmosphere vulnerable to erosion from large impacts.

Evidence for the loss of atmosphere comes not just from the worsening climate but also from the isotopes of the gases that remain in the air today. When the inert gas argon escapes into space, the light isotope argon-36 does so more readily than the heavier argon-38—and today the Martian ratio of argon-38 to argon-36 is 30 percent higher than on Earth. In contrast, argon trapped in the old Martian meteorite ALH 84001 resembles the terrestrial. This gas presumably preserves the old Martian atmosphere, when solar-wind stripping of the atmosphere had barely begun. Instead, at that time, impacts were devastating the atmosphere, and they remove all isotopes equally.

The isotope ratios of radiogenic noble gases—those produced by radioactive elements—lend further insight. Mars was born with three argon isotopes, not only argon-36 and argon-38 but also argon-40. All got blasted into space when asteroids and comets hit Mars. However, after the 800-million-year-long heavy bombardment let up, new argon-40 entered the atmosphere. That's because it, unlike the other argon isotopes, can arise when radioactive potassium-40, in the rocks below, decays; and potassium-40's half-life is a leisurely 1.3 billion years, so most of the radioactive potassium decayed after both hydrogen escape and the heavy bombardment had swept away the original mix of argon. As a result, modern Martian air has a high proportion of argon-40: the ratio of argon-40 to argon-36 is ten times higher than on Earth.

Ditto for xenon-129, which is over twice as common on Mars than Earth, relative to other xenon isotopes. Xenon-129 forms when radioactive iodine-129 decays. Hydrogen escape and large impacts removed most of the original xenon from the Martian atmosphere, so when the iodine-produced xenon-129 entered, it found itself more dominant relative to other xenon isotopes than on Earth. Iodine-129's half-life is only 17 million years, but the xenon produced by its decay took time to creep upward through the rocks and enter the air.

Nitrogen isotopes also show differences from Earth that indicate Mars has lost atmosphere. Most nitrogen on Earth, on Mars, and in the universe is nitrogen-14, the lighter variety; but some is nitrogen-15. Because of its lesser weight, nitrogen-14 rises higher and more readily enters the upper atmosphere, where it is vulnerable to escape. Over its life Mars has lost so much nitrogen that the ratio of nitrogen-15 to nitrogen-14 in its air is 70 percent greater than in Earth's.

Much of the ancient air that aided past Martian climate emerged from its volcanoes. As we see in FIRE, Martian volcanoes are colossal structures, far taller than any volcanoes on Earth or elsewhere in the solar system. Because of their precious progeny—the gases that warmed ancient Mars—they may hold the key to the climate that could have fostered Martian life.

FACING PAGE: The Sun's light warmed ancient Mars, but the solar wind helped tear away the same air which tried to retain that warmth.

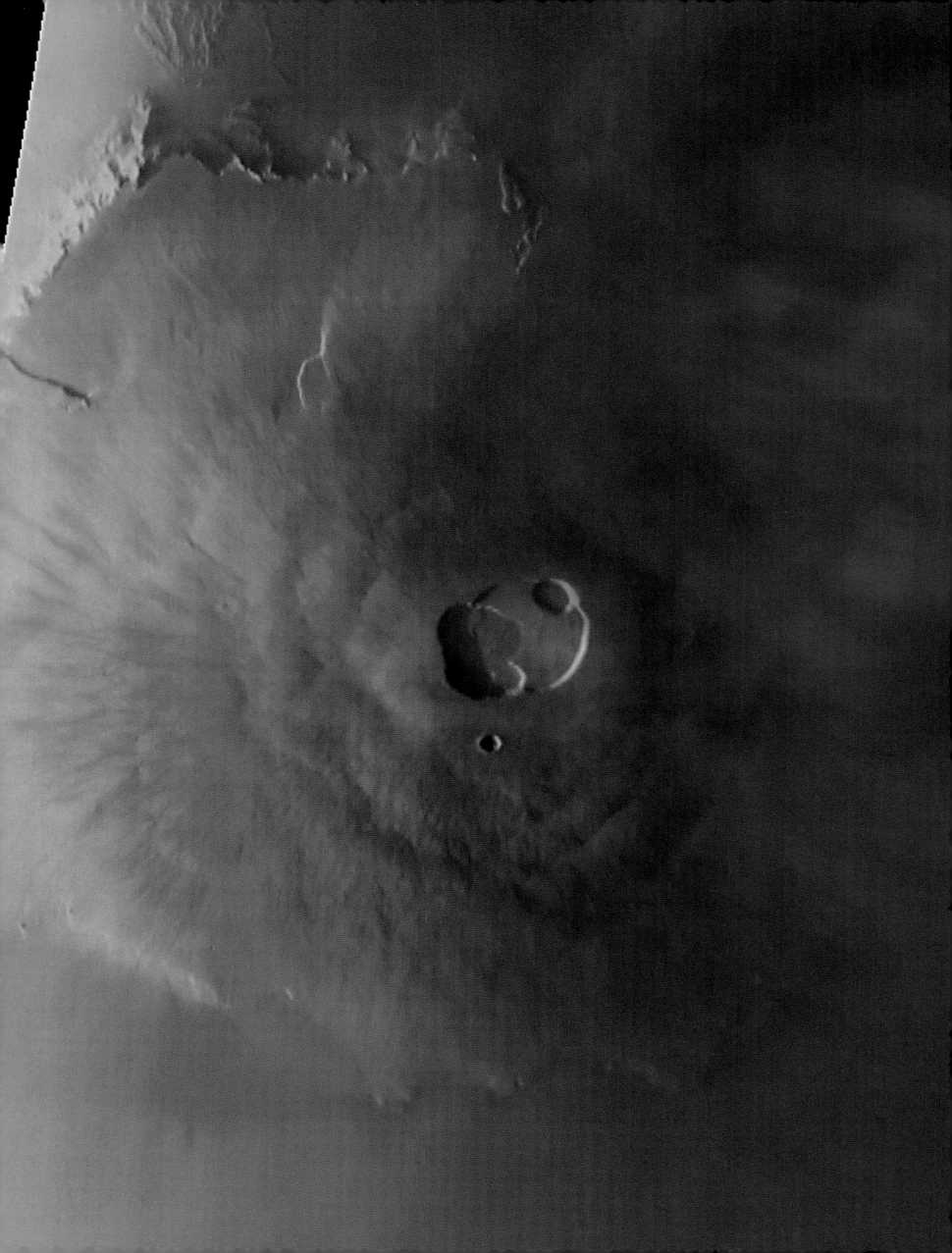

Fire

VOLCANOES BREATHE FIRE. Indeed, their name comes from the Roman fire god Vulcan. Although modern Mars exhibits less volcanic activity than Earth, Martian volcanoes vastly surpass their terrestrial peers in size. The red planet's tallest volcano—the appropriately named Olympus Mons—is over a hundred times more massive than the Earth's mightiest volcano. The same Martian plateau that spawned such enormous volcanoes also cracked the planet's crust, creating the solar system's largest canyons. Martian volcanoes have flooded large regions of the planet with lava and, in some cases, melted ice. On Earth volcanoes vent greenhouse gases that warm the world, and volcanoes likely did the same for ancient Mars, giving the planet a climate that may have fostered life.

FACING PAGE: Over twice as tall as Mount Everest, the Martian volcano Olympus Mons is the solar system's highest mountain. North is up.

Gentle Giants

LARGE though they are, the mightiest Martian volcanoes are gentle. They are called shield volcanoes, because their shallow slopes resemble the shape of a knight's shield. The most peaceful of all terrestrial volcanoes, shield volcanoes populate Hawaii and usually erupt nonexplosively. Hawaiian volcanoes rarely kill.

In contrast, just one eruption of a violent volcano can kill tens of thousands of people. The ferocity of a volcano stems largely from the composition of its magma. Geologists classify magma by its silica content. Silica is silicon dioxide (SiO_2). As described in EARTH, oxygen and silicon are the most abundant elements in the terrestrial and Martian crust. On Earth silica commonly manifests itself as the beautiful mineral quartz. Pure quartz is light-colored and glassy, but add some impurities and you have rose quartz or amethyst. Other varieties include agate, onyx, and jasper.

> ## Most of Mars is covered with volcanic rock, testimony to the power of Martian fire

With beauty, though, comes danger. The more silica the magma has, the more violent is its eruption. Silica slows the magma's flow, so the magma can clog the volcano. Then gases can build up beneath the clog and explode the volcano. Such an explosion of Mount Vesuvius destroyed the Italian cities of Pompeii, Herculaneum, and Stabiae in A.D. 79.

Magma comes in three main types. The safest, with the least silica, is called basaltic; it's what shield volcanoes erupt. Basaltic lava is about 50 percent sil-

ica by weight, and much of the rest is iron and magnesium oxides. Since iron and magnesium abound on the Martian surface, it's no surprise that the red planet has shield volcanoes. The low silica content facilitates the flow of basaltic lava, which is why shield volcanoes don't normally clog and explode. Indeed, the lava's fluidity explains not only the gentle eruptions but also the gentle slopes. Since the lava flows fast, even a gentle slope carries it away from the volcanic vent.

The other two magma types, with more silica, are called andesitic and rhyolitic. In andesitic lava, silica accounts for about 60 percent of the weight; in rhyolitic, about 70 percent. These lavas creep sluggishly, so volcanoes that erupt them can explode.

On Earth most lava is the safest variety: 80 percent of all terrestrial lava is basaltic. Andesitic and rhyolitic lava account for only 10 percent each. On Mars, with its high iron and magnesium content, basaltic lava is probably even more common. Indeed, many of the Martian meteorites are basaltic. However, the Sojourner rover that the Pathfinder spacecraft released found both basalt and andesite, and the Mars Global Surveyor's Thermal Emission Spectrometer revealed that the dark areas of the northern lowlands seem to be andesite, the dark areas of the southern highlands basalt. Most of Mars is covered with volcanic rock, testimony to the power of Martian fire.

Martian volcanoes congregate in three main regions. The mightiest reside on or near the Tharsis bulge. A smaller uplift, a third of the way around the planet from the Tharsis summit, has three volcanoes. This is Elysium, north of the equator. Volcanoes also stand far to the south, on the rim of the giant impact basin Hellas.

The Tharsis Volcanoes

THE GREATEST volcanic province on Mars, the Tharsis uplift dominates nearly half the planet. It splashes the topographic map with vivid hues of orange, red, and brown—designating high altitudes—and its tallest volcanoes are capped with topographic white. Tharsis and its surroundings house five great volcanoes, including the planet's three tallest, as well as numerous smaller volcanoes. Most Martian meteorites likely originated here.

An explorer starting west of the Tharsis bulge and proceeding east would first encounter the red planet's flagship volcano, Olympus Mons. Not only is Olympus Mons the tallest volcano on Mars, but it is also the tallest in the entire solar system. Named for the mythological mountain home of the Greek gods, Olympus Mons is over twice the height of Mount Everest. As mentioned in EARTH, however, the Mars Global Surveyor's laser altimeter reduced this mighty mountain's stature, finding its peak about 3 miles lower than had been thought. According to the new data, Olympus Mons is 21.287 kilometers, or 69,841 feet, above Martian sea level. For comparison, Earth's largest volcano, Mauna Loa in Hawaii, stands 9.1 kilometers above the Pacific seafloor and 4.2 kilometers above sea level. If Olympus Mons sat on the American East Coast, it could simultaneously smother Boston, New York, and Philadelphia. Yet like Mauna Loa, Olympus Mons is a shield volcano, with gentle slopes of just a few degrees.

However, a steep cliff, as high as 6 kilometers, surrounds the volcano. On the topographic map notice how the red high around Olympus Mons' peak quickly yields to the green low, with little or no intervening orange and yellow. North and west of Olympus Mons, colored green on the topographic map, is a strange terrain called the Olympus Mons aureole, made of lobes of ridges that run over hundreds of miles. Its origin is unknown.

Southeast of Olympus Mons, atop the topographic reds and oranges of the Tharsis bulge proper, tower three huge shield volcanoes, the Tharsis Montes. The three volcanoes form a diagonal line that straddles the equator from southwest to northeast. To a stargazer they recall the three stars in the belt of the constellation Orion. From southwest to northeast they are named Arsia Mons, Pavonis Mons, and Ascraeus Mons, after a forest northwest of Rome, the constellation Pavo the Peacock, and the birthplace of Greek poet Hesiod, respectively. None of the three quite achieves the heights of Olympus Mons, but all dwarf any mountain on Earth. These volcanoes may be less impressive than Olympus Mons because they have borrowed one another's magma, whereas Olympus Mons sits off to the side, hogging its magma for itself.

THE GREAT VOLCANOES OF MARS

VOLCANO	LOCATION	HEIGHT *		
		FEET	MILES	KILOMETERS
Olympus Mons	Near Tharsis	69,841	13.227	21.287
Ascraeus Mons	Tharsis	59,774	11.321	18.219
Arsia Mons	Tharsis	58,336	11.048	17.781
Elysium Mons	Elysium	46,347	8.778	14.127
Pavonis Mons	Tharsis	46,120	8.735	14.057
Alba Patera	Tharsis	22,211	4.207	6.770

*For comparison, Mount Everest is 29,035 feet = 5.499 miles = 8.850 kilometers above terrestrial sea level.

cano named Alba Patera that has no equivalent on Earth or elsewhere on Mars. *Alba* means the "white region"; *patera* means "saucer," and Alba Patera does not protrude the way its four great volcanic peers do. On the topographic map, it is a large reddish patch that never achieves the lofty browns and whites that mark the peaks of its peers. Alba Patera has not even a third the height of Olympus Mons. It is the only great Tharsis volcano that fails to surpass Mount Everest.

Yet in areal extent it outdoes all other Martian volcanoes, covering eight times the area of Olympus Mons and nearly two hundred times that of Earth's Mauna Loa. Perhaps Alba Patera owes its peculiarity to its high latitude. At such latitudes, the ground probably bears more ice; perhaps the lava melted so much ice that the mountain collapsed.

BELOW LEFT AND FACING PAGE: Tharsis and its surroundings support five great volcanoes: Olympus Mons, the Tharsis Montes (Arsia Mons, Pavonis Mons, and Ascraeus Mons), and Alba Patera. North is up.

BELOW RIGHT AND PAGES 102-103: The high altitudes of Tharsis color much of Mars topographic orange and red; the canyons of Valles Marineris, mostly topographic blue and green, run away from the Tharsis summit. North is up.

PAGES 104-105: Olympus Mons owes its immense size in part to the red planet's lack of continental drift. North is up.

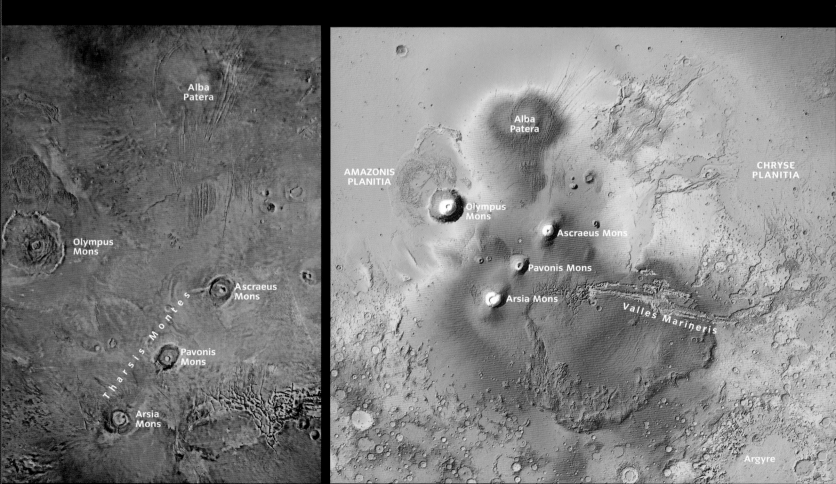

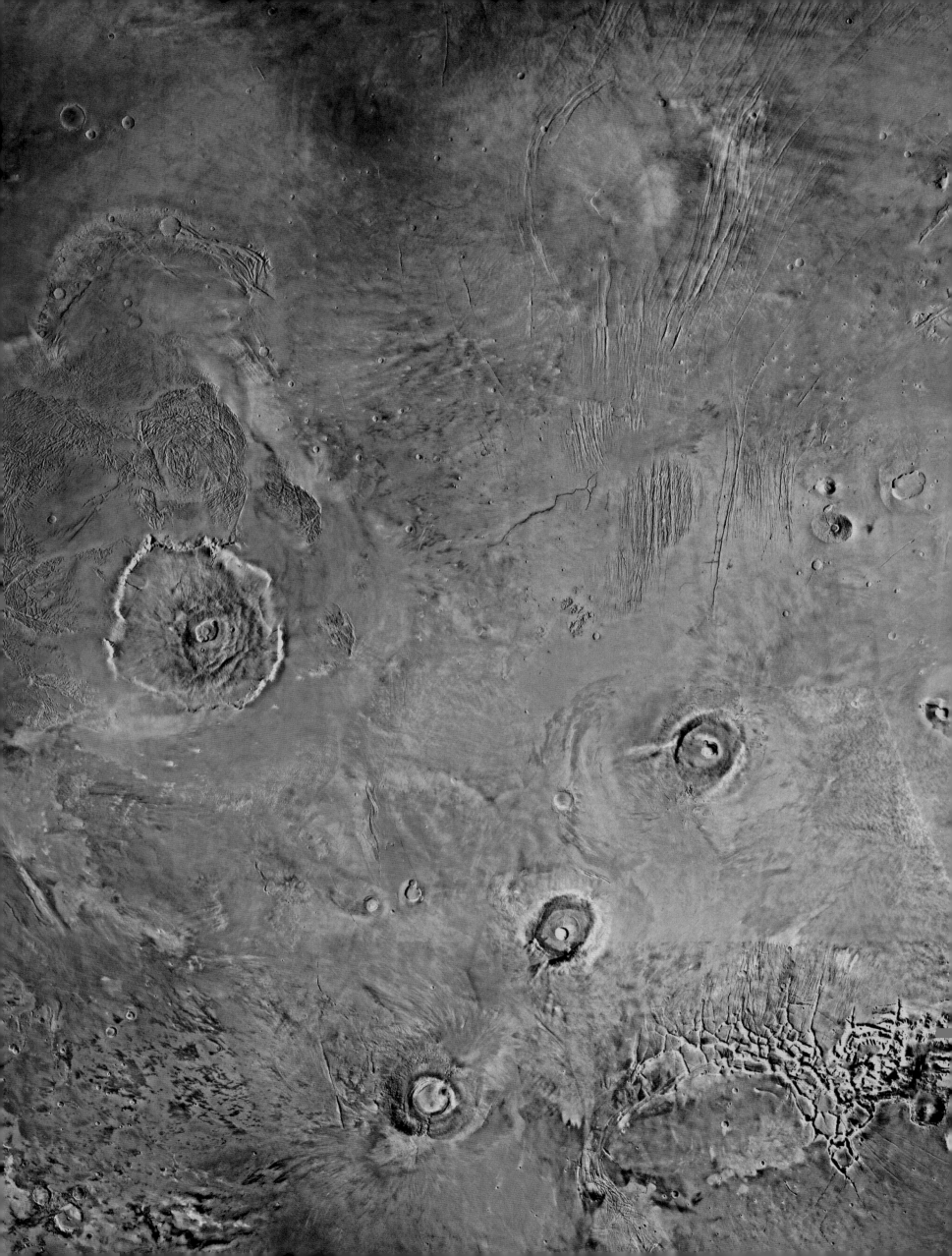

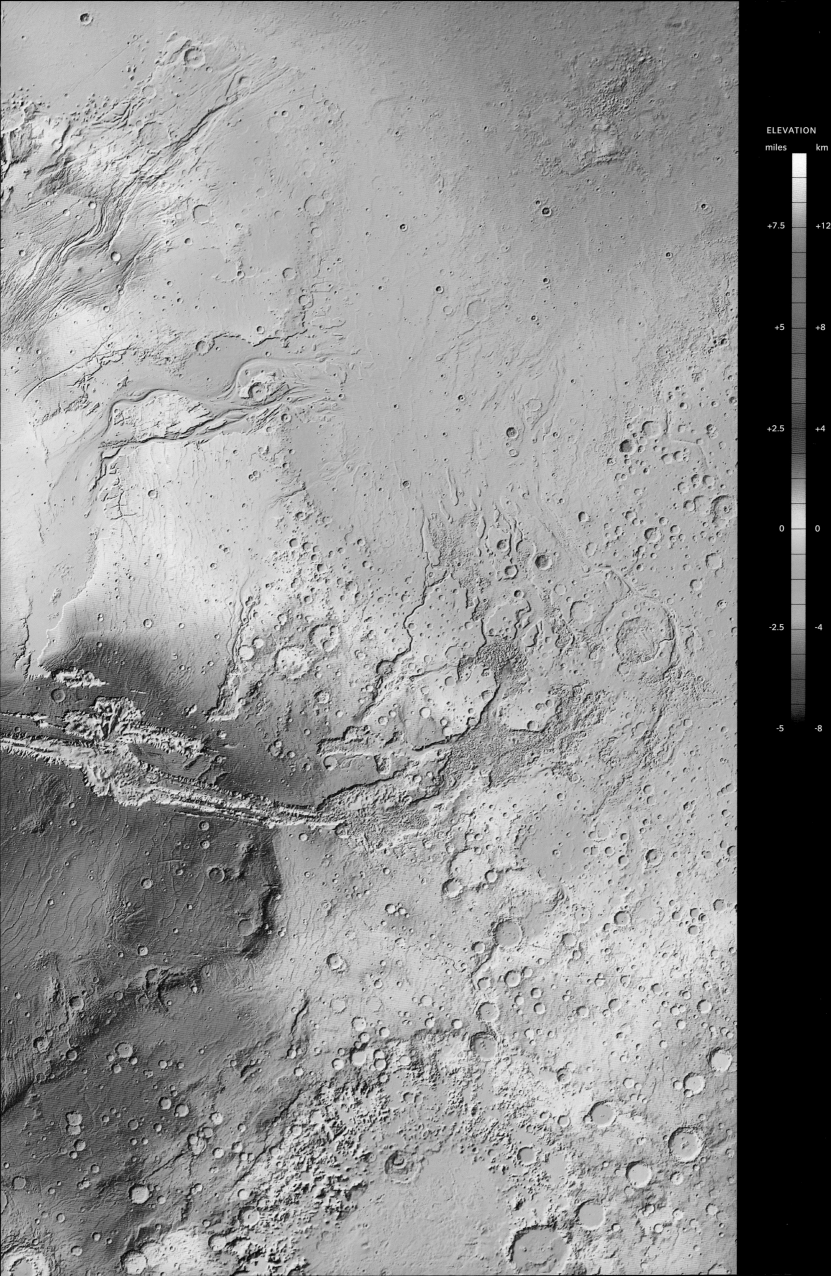

ELEVATION

miles	km
+7.5	+12
+5	+8
+2.5	+4
0	0
-2.5	-4
-5	-8

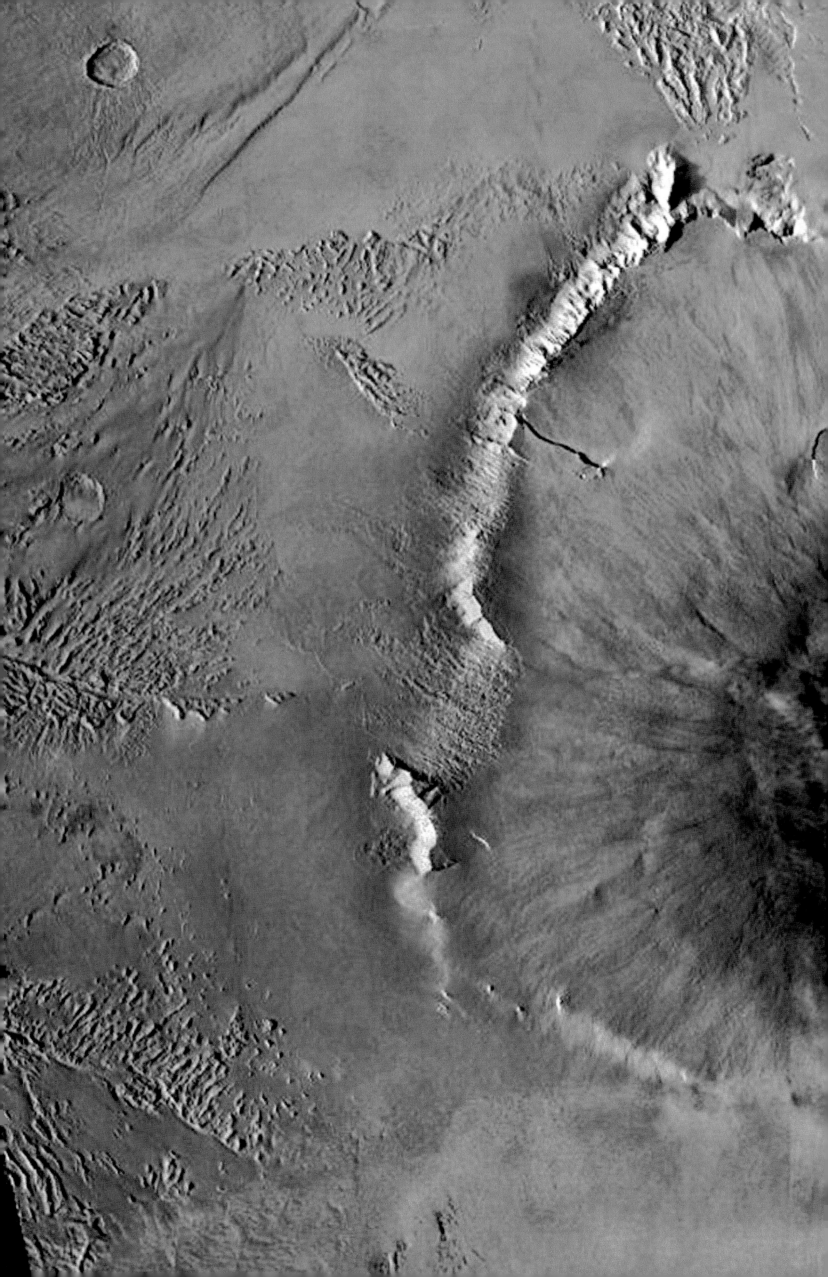

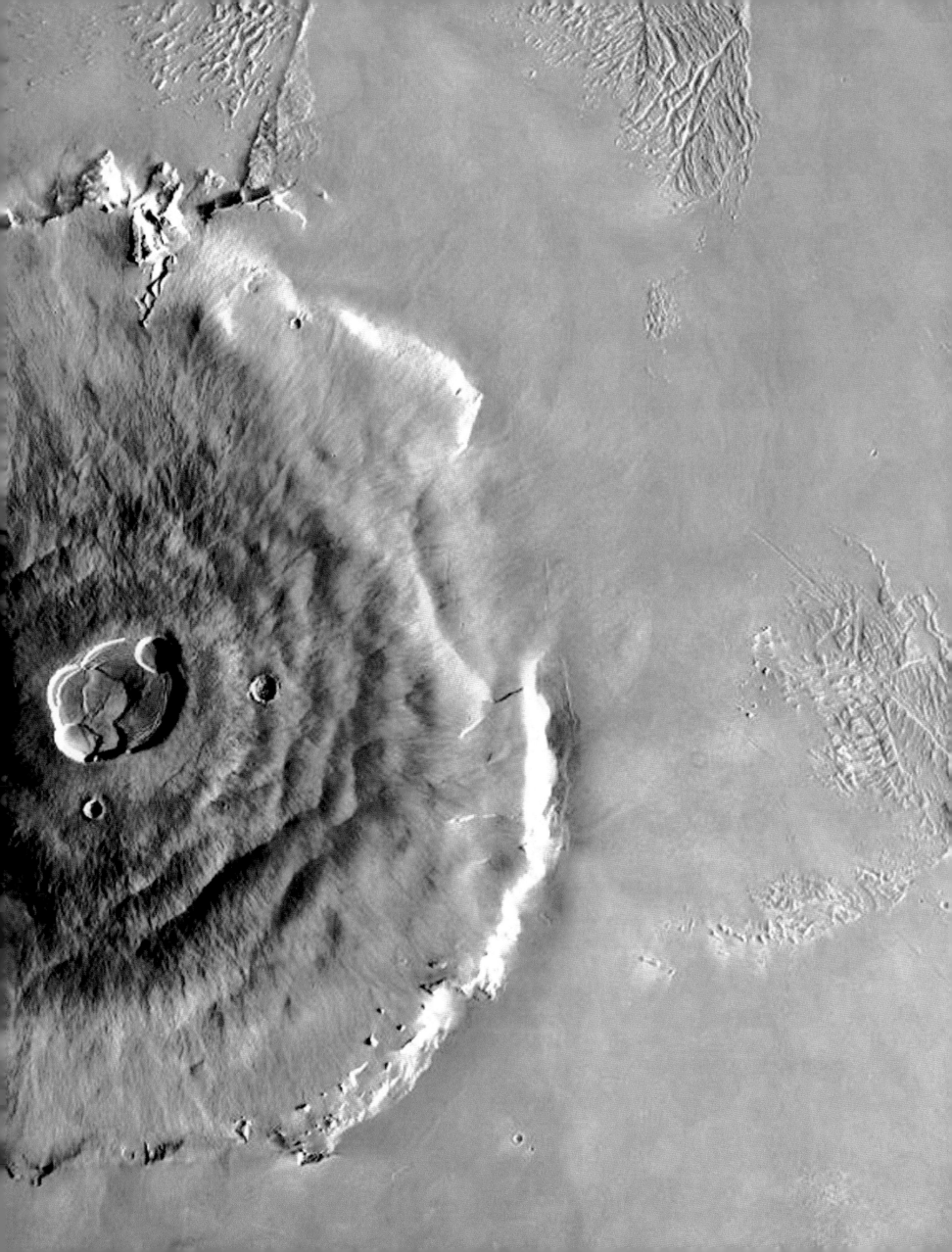

Aside from Alba Patera, the great Tharsis volcanoes resemble terrestrial shield volcanoes, except all are far larger. Not only are the volcanoes themselves huge, but so are their lava flows. Some run for hundreds of miles. According to a 2001 study, the Tharsis volcanoes erupted enough lava to bury the entire planet to a depth of 1.3 miles. No wonder most of the Martian surface is volcanic rock.

> The Tharsis volcanoes erupted enough lava to bury the entire planet to a depth of 1.3 miles. No wonder most of the Martian surface is volcanic rock

The volcanoes grew tall for three reasons. First, the thick Martian crust, described in EARTH, can support the enormous weight of a huge mountain. Earth's thinner crust can't. Second, the weaker gravity on Mars means that a Martian mountain weighs less than a terrestrial mountain of the same size. Third, Mars lacks continental drift, which limits the aspirations of terrestrial volcanoes. For example, the Hawaiian islands are volcanoes that owe their existence to a hot spot in the mantle below. Because of continental drift, however, the plate on which Hawaii rides—the Pacific plate—is moving northwestward, so any one island remains over the mantle's flame only briefly. During this time, the volcano erupts and builds the island. When it drifts away, the eruptions cease, so the volcano never achieves the Olympian heights of the greatest Martian volcanoes. Instead, the hot spot builds a new volcano. Because of this northwestward movement, the northwestern Hawaiian volcanoes are extinct, while the southeasternmost, Kilauea, is the most vigorous. Someday Kilauea will drift away from the hot spot and fall silent, and another island, not yet born, will take its place in the volcanic limelight. Geologists estimate that a Hawaiian volcano remains active for only a few hundred thousand years. In contrast, a Martian volcano like Olympus Mons can sit over a mantle hot spot for billions of years and achieve immense heights.

Not all Tharsis volcanoes prospered, however. Numerous smaller volcanoes, such as Uranius Patera and Tharsis Tholus, cower in the shadows of the giants. Strangely, no volcanoes at all occupy the southeastern half of Tharsis. Might Tharsis be the inverse of Hawaii, its youngest volcanoes to the northwest, and its oldest, in the southeast, eroded out of existence?

When the Mars Global Surveyor's laser altimeter measured the Tharsis topography, it found that Tharsis differed from previous expectations. Rather than a single bulge, Tharsis proved to be a topographic barbell made of two bulges—a small one to the north, beneath the peculiar volcano Alba Patera, and a much larger one to the south, on whose north-western edge stand the three Tharsis Montes. The greatest Tharsis volcano, Olympus Mons, does not actually reside on the Tharsis uplift but instead lies northwest of it. On the topographic map the other great Tharsis volcanoes sit on land colored red and orange, whereas the land surrounding Olympus Mons is yellow and green.

The Tharsis uplift formed early in the red planet's life. Scientists know this because the Mars Global Surveyor's laser altimeter found that ancient rivers in and around Tharsis followed the current topography, flowing downhill. As described in WATER, most of these rivers flowed during the Noachian period, so the Tharsis uplift must have been in place during the oldest period of Martian history. Old lava flows also conform to the current topography. The sur-faces in Tharsis look younger—mostly of Amazonian and Hesperian age—because the volcanoes have spewed lava over impact craters. Tharsis probably formed when a plume of hot mantle material rose, pushing the crust upward and stimulating volcanic eruptions, whose lava augmented the rise.

Tharsis is centered on the Martian equator, which may or may not be a coincidence. A planet with a bulge tries to reorient itself so that the equator bears the burden. Thus, either Tharsis happened to arise on the equator or it formed elsewhere and the planet pirouetted. If the latter, then Mars once had dif-ferent poles. Since Tharsis formed long ago, however, any reorientation also occurred long ago, making it difficult to locate a record of past poles on the planet's present surface.

Some Tharsis volcanoes probably still erupt. After all, most Martian meteorites are young volcanic rocks—with ages from 165 million to 1.3 billion years—and though no one knows where on Mars they came from, the best bet is Tharsis. And if the Tharsis volcanoes erupted a mere 165 million years ago, just 4 percent of the red planet's age, they probably do so today. Also, images from Mars Global Surveyor, which William K. Hartmann of the Planetary Science Institute in Tucson, Arizona, and his colleagues analyzed in 1998, show that many lava flows in Tharsis and else-where bear few impact craters, indicating that the lava flowed recently. For example, Arsia Mons erupted just 40 million to 100 million years ago, and Olympus Mons has fresh lava flows that formed within just the past 20 million years—less than half a percent of the planet's age.

The Tharsis uplift formed early in the red planet's life

The Elysium Volcanoes

ALTHOUGH THARSIS houses the most famous Martian volcanoes, it is not the only keeper of the red planet's flame. A small uplift on the planet's other side also has volcanoes. This uplift is named Elysium, which Homer's *The Odyssey* called a place "where life is easiest. No snow is there, no heavy storm, nor ever rain." The Elysium volcanoes are smaller than the greatest in Tharsis, but the largest peak still outclasses any mountain on Earth. Like Tharsis, the Elysium uplift probably formed from a hot plume of rising mantle material, presumably smaller than the one that raised Tharsis.

Elysium has three volcanoes. They reside north of the equator—they have a Hawaiian latitude—in a roughly north-south line. The largest, Elysium Mons, is the central peak; it is the fourth tallest volcano on Mars overall. Two smaller volcanoes, Hecates Tholus to the north and Albor Tholus to the south, round out the Elysium complex. Over three dozen sinuous rilles—collapsed lava tubes—run over Elysium, some exceeding a hundred miles in length. Long channels west of Elysium Mons suggest that the volcanoes melted ice and caused water to flow into neighboring Utopia, the northern impact basin where Viking 2 landed. Could some of the many rocks that Viking 2 saw have originated in Elysium?

At least one Elysium volcano probably still erupts. Mars Global Surveyor images show barely any impact craters scarring parts of southern Elysium Planitia, the volcanic plain south of Elysium Mons and its peers. Indeed, the youngest lava flows here have less than 1 percent as many craters per area as the smooth lunar seas. From the smoothness of the Elysium lava flows, William K. Hartmann and Daniel Berman of the Planetary Science Institute in 2000 concluded that the lava is younger than 100 million years, and some may be younger than 10 million years—just 0.2 percent of the planet's age. To an eighty-year-old, that's the equivalent of only two months ago.

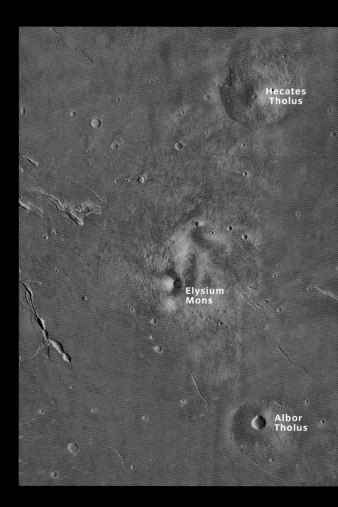

ABOVE AND FACING PAGE: The Elysium rise has three volcanoes. From north to south, they are Hecates Tholus, Elysium Mons, and Albor Tholus. North is up.

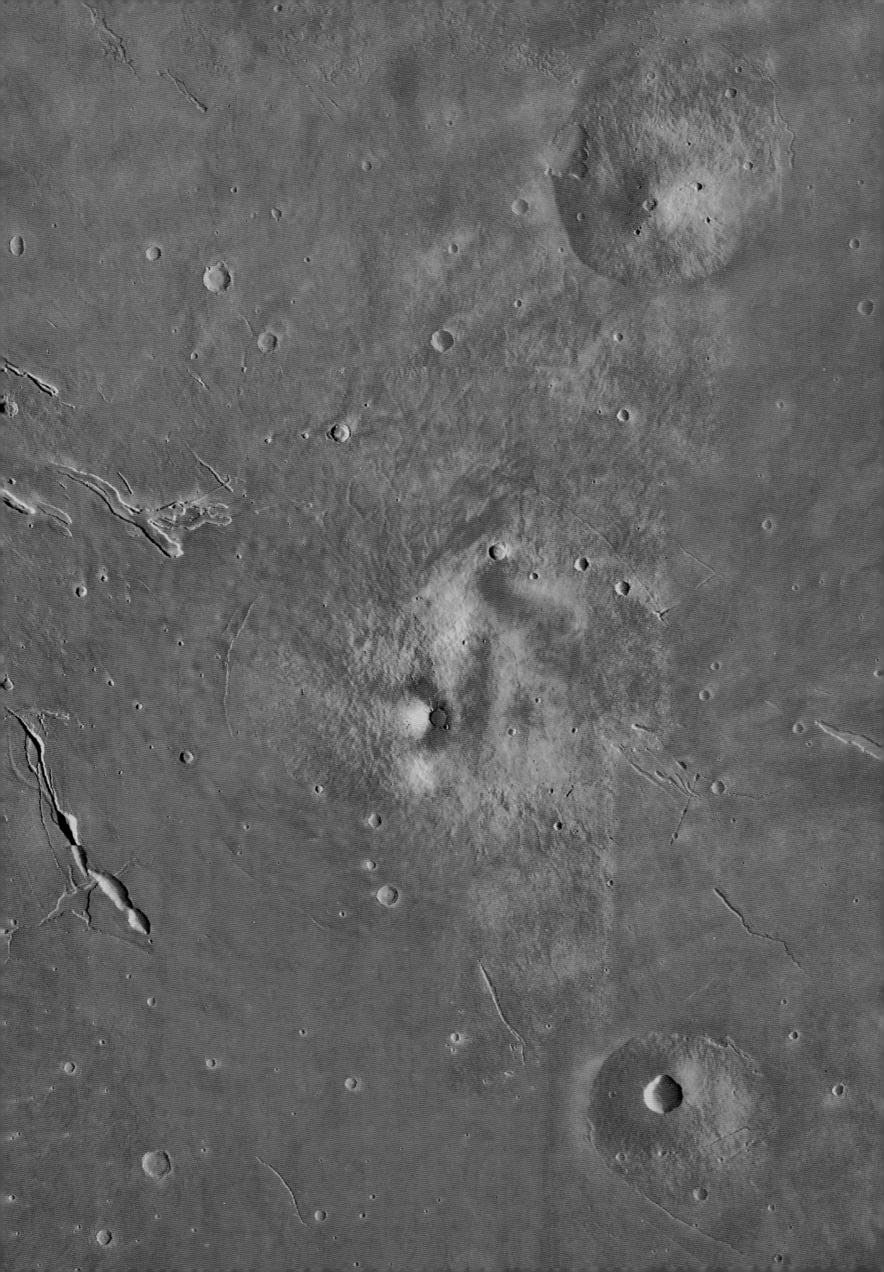

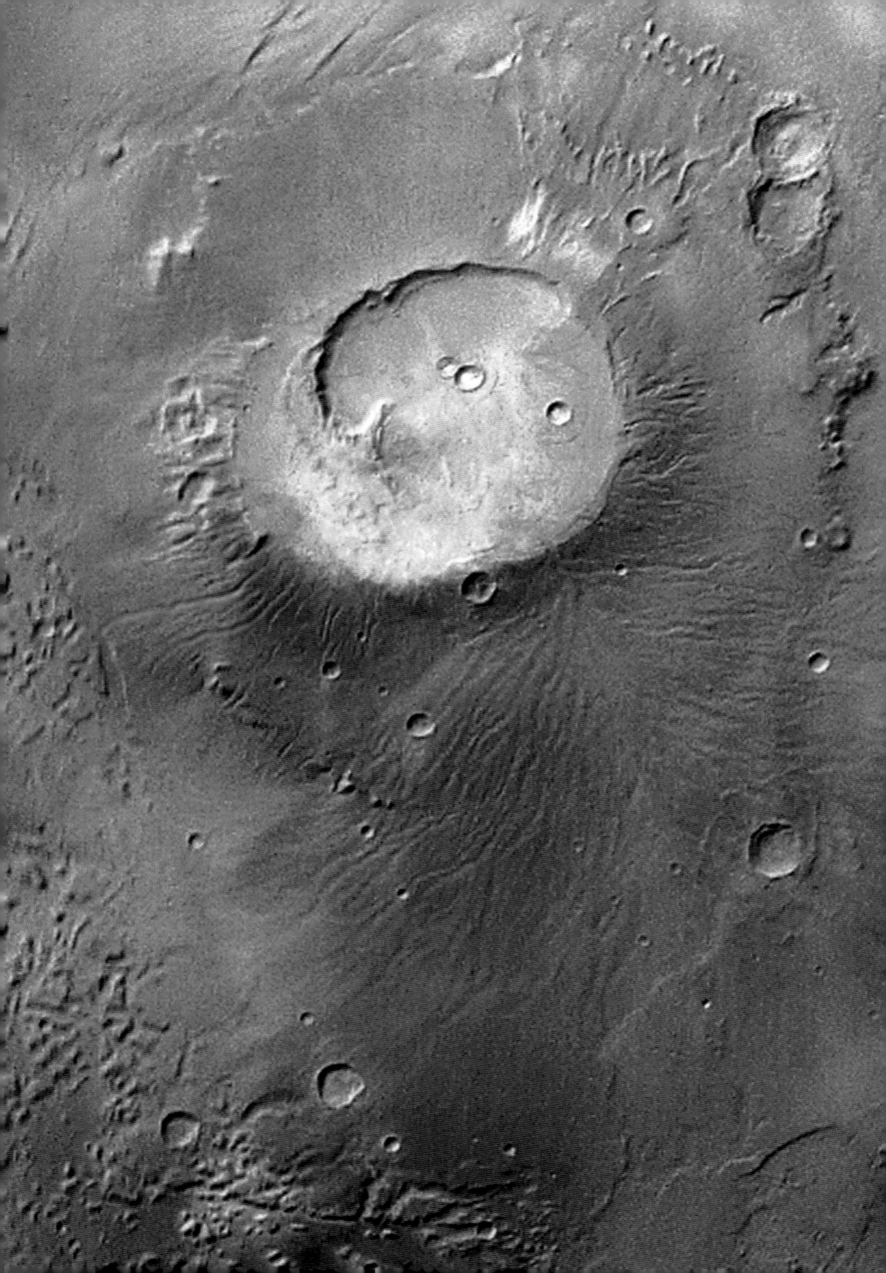

The Hellas Volcanoes

HELLAS, the giant basin in southern Mars, formed when a large asteroid smashed into the young planet. The impact cracked the surface, leaving fissures through which underground magma may have erupted. Two volcanoes, named Hadriaca Patera and Tyrrhena Patera, reside on Hellas's northeastern rim, while two others, named Amphitrites Patera and Peneus Patera, stand on the rim opposite, in the southwest. Unlike many other Martian volcanoes, the Hellas volcanoes may have erupted explosively. Possible ash deposits surround Tyrrhena Patera, and ash—as opposed to lava—signifies volcanic violence.

The culprit may have been water. If water invades its vent, even an otherwise peaceful volcano can erupt explosively. This happened in Hawaii in 1924, when water seeped into Kilauea and the subsequent explosive eruption killed one person. As described in WATER, the Hellas volcanism may have melted ice and sent water flowing into Hellas. Two large valleys east of Hellas drain into the basin, and Hellas itself may have once hosted a large lake. Yet the Hellas volcanoes may also testify to the drying up of the planet. Lava flows over, and therefore is younger than, the possible ash deposits, suggesting that the water supply dwindled, causing the eruptions to switch from violent to gentle.

For some time scientists have noticed that Hellas lies on the opposite side of the planet from Tharsis. On Earth the asteroid that killed off the dinosaurs struck Mexico, which is approximately opposite India, where volcanoes erupted around the same time—leading some scientists to think the volcanic eruptions, not the impact, doomed the dinosaurs. Some have even suggested that shock waves from the impact sparked the volcanism. On Mars could the Hellas impact have done the same in Tharsis? Probably not. First, Hellas is not directly opposite Tharsis, although if Mars once had continental drift, Tharsis may have moved. Second, it's hard to see how the impact could stimulate so much volcanism on the opposite side of the planet when it produced so little near itself. On Mercury, a large asteroid carved out a basin called Caloris, whose name means "heat," appropriate for the Sun's innermost planet; on the opposite side of Mercury from Caloris Basin is not a volcanic province but instead hilly terrain, probably produced when shock waves from the impact raced around the planet and converged on the opposite side.

Not all Martian volcanoes reside in Tharsis, Elysium, or Hellas. For example, over a thousand miles southeast of the Elysium volcanoes stands the isolated volcano Apollinaris Patera. It's just south of the equator, near the boundary between the northern lowlands and the southern highlands. Elsewhere on Mars, domes and hills may be volcanoes, too. In 2000, Jim Garvin of NASA's Goddard Space Flight Center in Greenbelt, Maryland, and his colleagues used the topographic data from the Mars Global Surveyor's laser altimeter to locate several possible shield volcanoes in the far north of Mars. They may be only half a million to 20 million years old.

FACING PAGE: White clouds cling to the summit of Apollinaris Patera, an isolated volcano just south of the Martian equator. North is up.

Valles Marineris

WHEN THE THARSIS plateau arose on Mars, its uplift split the Martian crust, creating the largest canyon in the solar system. Named for the Mariner 9 spacecraft that discovered it, Valles Marineris is really a series of canyons, in eastern Tharsis. On the topographic map these canyons appear as green and blue lines slicing nearly east-west through red and orange terrain. Notice how they run almost directly away from the brown Tharsis summit.

West of Valles Marineris, near the Tharsis summit, fractured terrain called Noctis Labyrinthus leads into the canyons proper—on the topographic map the fractures are incisions in brown terrain. The canyons end far to the east, in turquoise terrain that feeds into several flood channels. These flood channels, also turquoise, drain northward into Chryse Planitia, where Viking 1 and Pathfinder landed. The flood channels

suggest either that the creation of Valles Marineris released enormous quantities of groundwater or that the canyons housed lakes that later spilled into Chryse.

The Valles Marineris system runs 2,500 miles, nearly the distance from London to Tehran. At its widest it consists of three connected canyons that together span nearly 400 miles; the deepest section is 6 miles beneath the adjoining cliff. Valles Marineris easily dwarfs Arizona's Grand Canyon, which is 277 miles long, at most 18 miles wide, and about a mile deep. Whereas a river carved the Grand Canyon,

BELOW AND PAGES I I 4- I I 5: Stretching 2,500 miles, Valles Marineris is the longest canyon in the solar system. North is up.

PAGES I I 6- I I 7: Enormous landslides bury the floor of the northernmost canyon shown here, Ophir Chasma. Some cliffs reach six miles down. North is up.

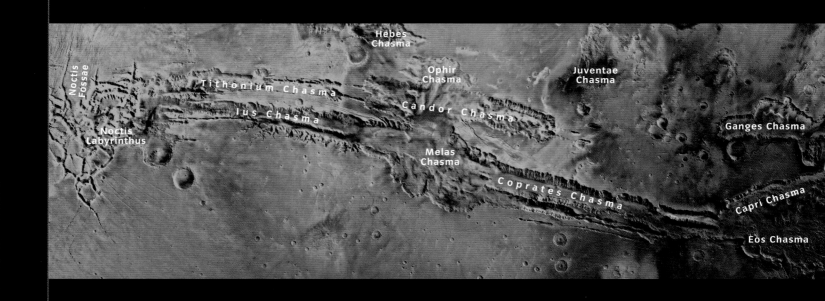

Valles Marineris owes its existence to the Tharsis uplift; however, floods rushing down the Martian canyons may have widened and deepened them. The canyons also sport enormous landslides, some over 60 miles wide. The canyon walls have strata that presumably preserve a record of Martian history. The strata may be deposits from ancient lakes that assumed different levels in the canyons at different times, or they may instead be lava flows.

Although Valles Marineris is the most dramatic example, other faults exist elsewhere in Tharsis. For example, long north-south grooves, visible on the topographic map, run for hundreds of miles near the odd Tharsis volcano Alba Patera, the one far to the north of the other great Tharsis volcanoes. Cracks also appear around the Elysium uplift.

Floods rushing down the Martian canyons may have widened and deepened them

Marsquakes

WHEN THE THARSIS uplift ripped open the Martian crust, the planet must have suffered gargantuan quakes. Modern Mars, though, is quieter, as the Viking landers showed. Viking 1's seismometer failed to work, but the Viking 2 seismometer operated for 3.5 years and detected nothing, apart from one event that was probably just the rustling of the wind. Viking 2 could best detect marsquakes near itself, in Utopia, but that is far from the volcanoes of Tharsis, where occasional eruptions may trigger quakes. The spacecraft would have missed a magnitude 8 marsquake on the opposite side of the planet.

Still, Mars is probably the perfect planet for quakephobes. It lacks the continental drift that spawns most terrestrial quakes, such as those rocking California and Japan. Furthermore, as noted in AIR, the Martian atmosphere has little helium and argon, two gases that arise in the planet's interior when radioactive elements decay. On Earth, quakes, volcanic eruptions, and continental drift shake these gases loose and inject them into the atmosphere, so their rarity in the Martian air again suggests a world much quieter than Earth.

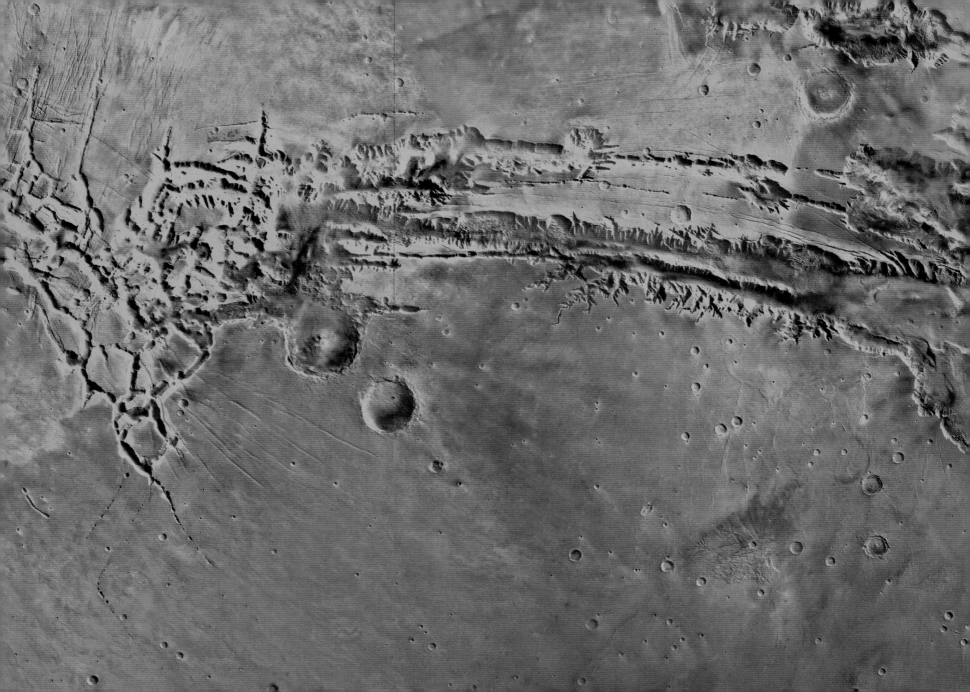

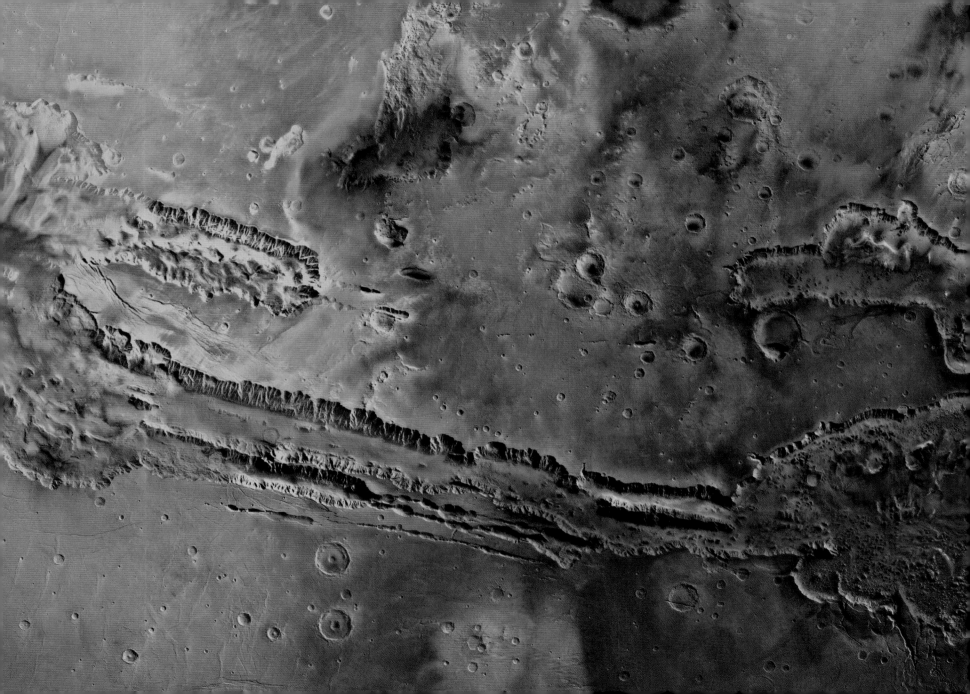

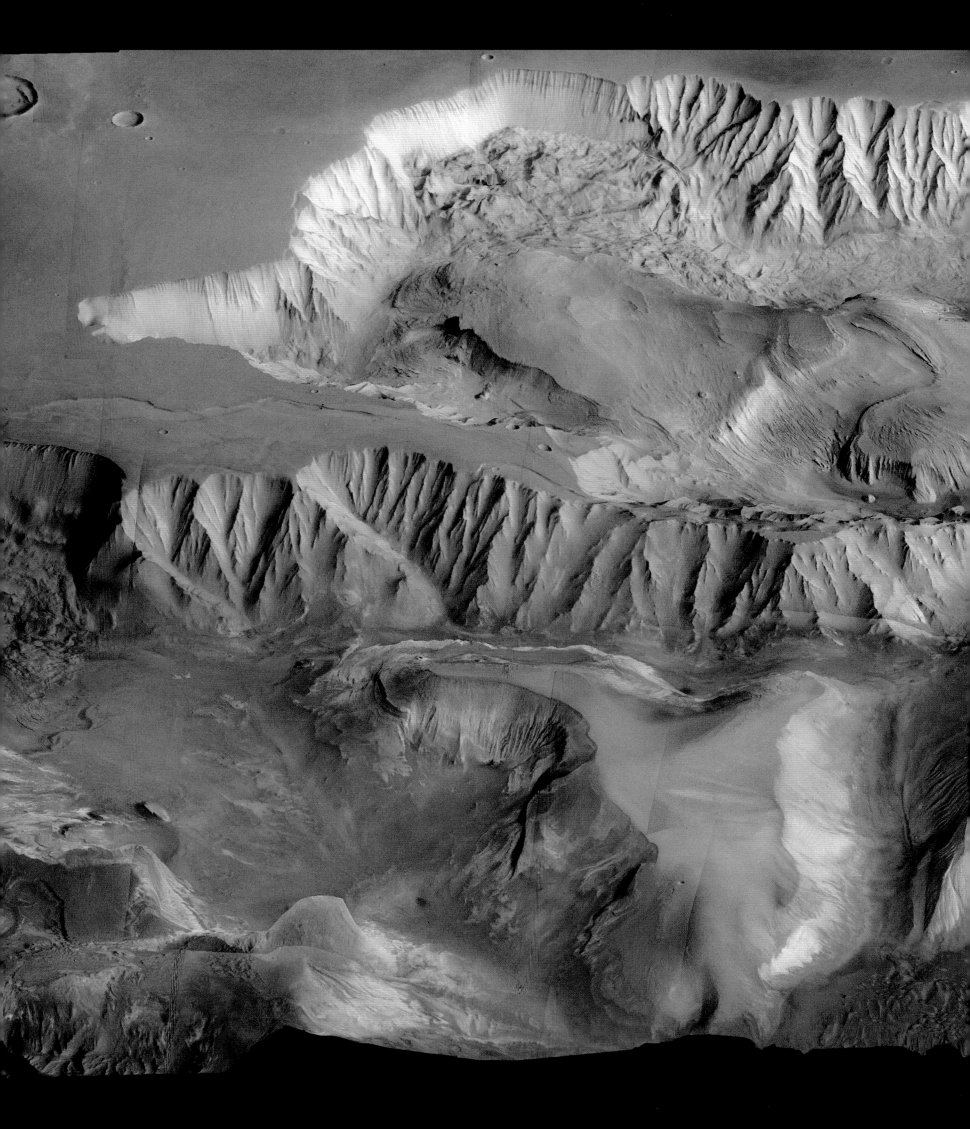

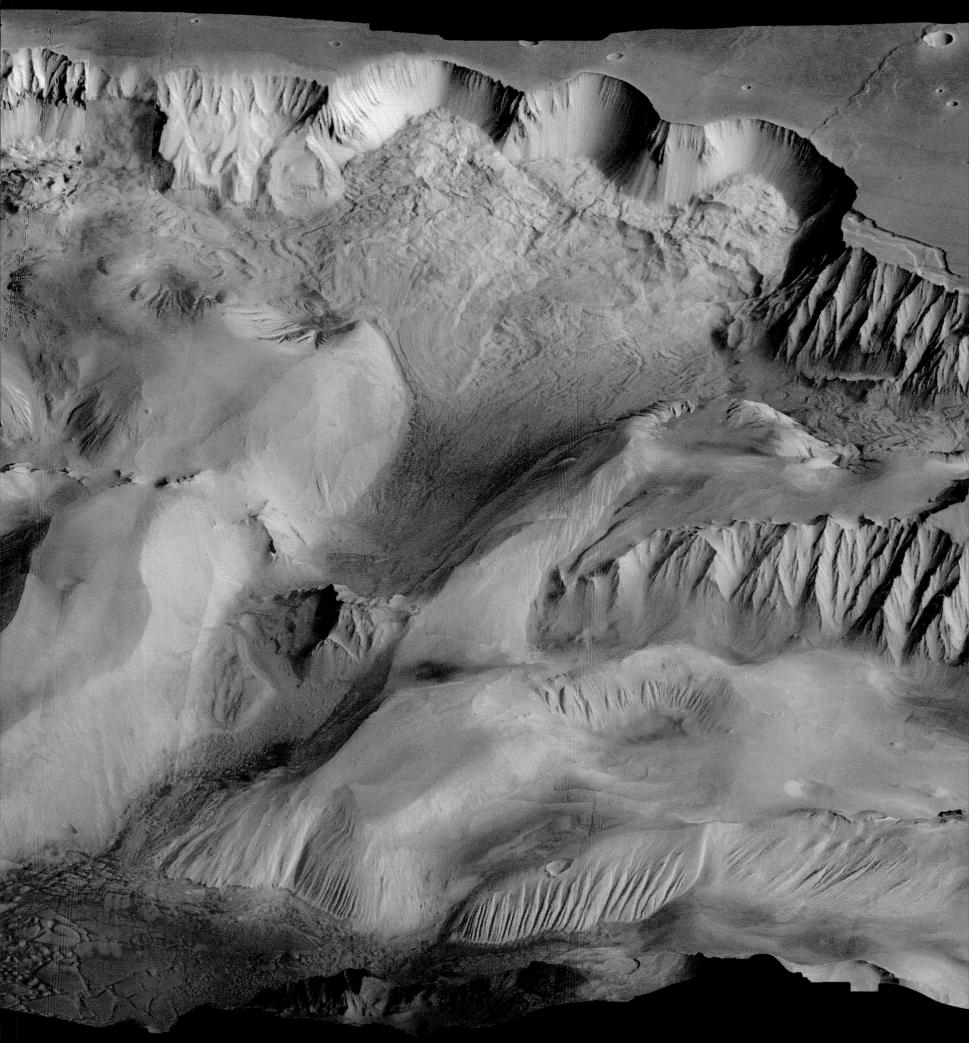

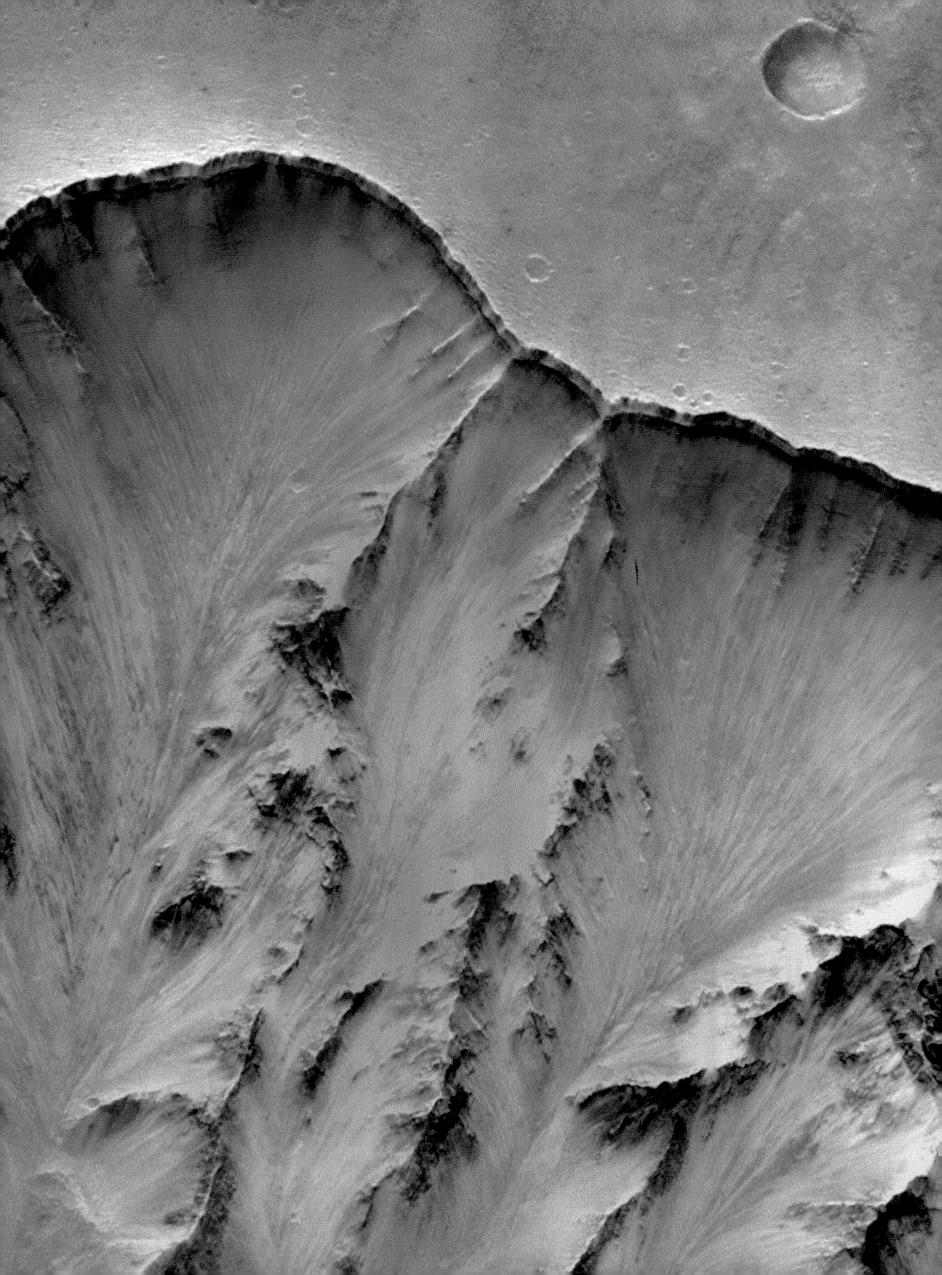

Martian Volcanoes Through the Ages

BY BREATHING FIRE—and air—the Martian volcanoes helped set the stage for the possible development of life on Mars. Early in the planet's life the Tharsis region arose, its volcanoes spewing lava over the surface and injecting gases into the air. The carbon dioxide, water vapor, and sulfur dioxide they emitted blessed Mars with an atmosphere similar in composition, if not in pressure, to the Earth's early atmosphere. Furthermore, these greenhouse gases helped warm this rather distant planet at a time when the Sun shone about 30 percent more faintly than it does now. So great was Tharsis volcanism, scientists in 2001 estimated, that it could have produced a carbon dioxide atmosphere with 1.5 times the pressure of the present terrestrial atmosphere. They also estimated that the Tharsis volcanoes vented so much water vapor that, if converted to liquid and spread evenly over the planet, it would have reached a depth of 120 meters, or 400 feet. In addition, the volcanoes released sulfur dioxide, a greenhouse gas even more potent than carbon dioxide and water vapor. Perhaps the volcanoes explain the high sulfur abundance that the Viking landers discovered in the red planet's soil.

Volcanism in Tharsis and elsewhere, then, may have warmed Mars so that it enjoyed a brief taste of Eden, when rivers flowed, lakes filled, and perhaps even an ocean weighed down the northern plains. Somewhere, a warm pool of sunlit water near a vol-

> Volcanism may have warmed Mars so that it enjoyed a brief taste of Eden

cano may have spawned the first life in the solar system—possibly predating that on Earth. Fossils of this ancient life may still exist on Mars today.

Any mild Mars, however, met with disaster. The volcanoes largely shut off, the magnetic field vanished, impacts and the solar wind tore the air away, and the planet's surface froze. As WATER describes, this climate change manifests itself most vividly in the story of the rivers and possible lakes that speckled the red planet's youth.

FACING PAGE: This close-up shows the northern cliffs of a small part of eastern Coprates Chasma. North is up.

Water

WATER! A key ingredient of all terrestrial life, water pours from the Earth's storm clouds, rushes down its rivers, and washes the continental shoreline. Water once flowed on Mars, too. Enormous floods, far greater than any known on Earth, inundated vast expanses north and northeast of the Valles Marineris canyons, carving flood channels into the red planet's terrain. Water flowed more peacefully, too, in rivers that ran through the cratered highlands. Lakes may have filled the Hellas and Argyre impact basins, and the northern lowlands may have even hosted an ocean. Some of that water still survives today, frozen at the icy poles or beneath the frigid surface. Billions of years ago, on a world warmer and wetter, that water may have sparked the first Martians, who left fossils documenting their struggle to survive in a world turning against them.

FACING PAGE: Water likely carved Scamander Vallis, a Martian valley in Arabia Terra.

be life. All terrestrial creatures, even those withstanding deadly gases or boiling temperatures, need water. Life probably arose in water, and even people who never see the ocean carry it with them: in most people, water accounts for most of their weight.

Water works wonders. It is an excellent solvent, dissolving various substances and providing a medium in which they can interact. Unlike most liquids, water freezes from the top down, so creatures living on the bottom of an ice-covered lake continue to bathe in liquid. Water exhibits this useful property because when it freezes it expands. As a result, its ice is lighter than its liquid and floats to the top. In con-

> The Earth is special because its warmth keeps its water liquid. Nearly 98 percent of terrestrial water is liquid

trast, most other ices sink in their own liquid.

Water abounds throughout much of the solar system and indeed the entire universe. It joins the universe's two most abundant reactive elements, hydrogen and oxygen (H_2O). The hydrogen arose in the big bang that set the cosmos expanding, the oxygen in massive stars that cast the element into space when they exploded. Water is common in the outer solar system. The rings of Saturn contain water ice, as do most of the moons of Jupiter, Saturn, Uranus, Neptune, and Pluto. Comets carry water ice, too.

Frozen water doesn't do much good, however. The Earth is special because its warmth keeps its water liquid. Nearly 98 percent of terrestrial water is liquid, a bit over 2 percent is ice, and 0.001 percent is gas. The importance of liquid water stimulates interest in Jupiter's moon Europa, whose icy surface may conceal an ocean that Jovian tides warm.

Mars has water. Trouble is, it's so cold that any on the surface is frozen. Furthermore, because the red planet's air is dry and thin, Martian ice behaves differently than terrestrial. At frigid temperatures—below about −105 degrees F.—ice remains ice, neither melting nor vaporizing. Above this temperature ice slowly vaporizes, going straight from ice to gas, with none of the intervening liquid that Earthlings take for granted. Ice behaves this way even above 32 degrees F.: if you tossed some ice onto a warm part of Mars—say, at 80 degrees F.—it would vaporize. If you tossed some liquid water onto this warm part of Mars, it would, strangely enough, *freeze*. That's because some of the water would rapidly evaporate in the dry, thin air; this rapid evaporation would so chill the remaining liquid that it would turn to ice; then the ice would slowly vaporize.

Astronomers first detected the spectral signature of Martian water in 1963, two years before the first Mars flyby, following a technique that canal champion Percival Lowell had suggested. Like other substances, water absorbs certain wavelengths of light, so Lowell tried to discern these absorptions in the Martian spectrum. Unfortunately, the Earth's moist atmosphere interfered: how to tell whether an absorption line arose from water on Mars or merely from terrestrial water vapor the telescope was forced to peer through?

To circumvent this problem, Lowell tried to exploit the Doppler shift. Because the Earth and Mars race around the Sun, sometimes Mars moves toward the Earth and sometimes away. When Mars moves toward us, its wavelengths get scrunched up, producing a blueshift, so called because blue light waves are short; when Mars moves away, the wavelengths get stretched out, producing a redshift. These blueshifts and redshifts, thought Lowell, could separate the Martian spectral lines from the terrestrial, helping him find Martian water.

Sound though Lowell's idea was, the technology of his day wasn't up to it. That had to wait for Hyron Spinrad, Guido Münch, and Lewis Kaplan, who used the 100-inch telescope atop California's Mount Wilson. By taking advantage of the red planet's redshift, they picked up atmospheric water vapor with a spectral wavelength around 8,200 angstroms, beyond the red end of the visible spectrum, in the infrared.

On Earth water does more than nourish life. It also erodes rock and carries debris from land to sea. Each year terrestrial rivers transport 14 billion tons of sediment into the ocean—that's over 2 tons per person. Water is thus a topographic socialist, lowering the heights and raising the depths. In part because modern Mars lacks flowing water, it exhibits far greater topographic diversity—taller mountains and deeper basins. For the same reason, Mars has better maintained its ancient surface, preserving a record of the

Mars has better maintained its ancient surface, preserving a record of the conditions under which Martian life may have struggled to arise, whereas the very water that sustains terrestrial life helped wash away signs of its birth

conditions under which Martian life may have struggled to arise, whereas the very water that sustains terrestrial life helped wash away signs of its birth.

planet's water comes from its floods. North and northeast of Valles Marineris several flood channels—Kasei Vallis, Shalbatana Vallis, Simud Vallis, Tiu Vallis, and Ares Vallis—spill into Chryse Planitia, the northern plain where the Viking 1 and Pathfinder spacecraft landed. On the topographic maps most of these flood channels are blue swaths hundreds of miles long cutting through green terrain. The channels run well north of both spacecraft, up to at least 30 degrees latitude.

The flood channels sport islands sculpted into teardrops: the water struck the round side of the teardrop and pulled out a long tail downstream. Grooves follow the channels and diverge around craters and other obstacles, just as water does. These features were not created by rivers. Rivers meander, they take their time, they tend to be narrow. Floods, on the other hand, have only one thing in mind: getting to low ground as fast as possible. They barely bend, they rarely have tributaries, they're wide but not deep, and they dwarf mere rivers. Some of the Martian flood channels are over a thousand miles long.

Except for their greater size, the Martian floods resembled the largest known terrestrial floods, which inundated the Pacific Northwest some 14,000 years ago. During the last ice age a large lake, Lake Missoula, occupied western Montana, holding half as much water as present-day Lake Michigan. Glaciers partially surrounded this lake. As the climate warmed, however, the ice burst, releasing enormous quantities of water westward, over Washington and Oregon, out to the Pacific Ocean—a journey of some 500 miles—burying the sites of Spokane and Portland beneath floodwaters hundreds of feet deep. Like the Martian

floods, these floods cut channels, now called the Channeled Scablands, in which teardrop-shaped islands testify to the flood's ferocity. Each flood lasted a few days and at its peak discharged dozens of times more water than the Amazon. Yet the Martian floods were larger still. Most of the Martian flood channels are of Late Hesperian age, but some are more recent, being of Amazonian age.

What triggered the floods? Some of the channels that empty into Chryse emerge from irregular, so-called chaotic terrain. The land here may have collapsed, as if something hit it and released enormous stores of water beneath. Perhaps a large impact did the trick, heating subsurface ice. Or perhaps a huge marsquake disrupted lakes in nearby Valles Marineris, causing their water to gush out. Elsewhere, volcanic eruptions may have melted subsurface ice and flooded Martian land with water.

Dramatic though they were, the floods do not prove that Mars was once warmer. So much water poured forth so fast that the water would not have frozen, even in the present frigid climate, until after the floods ended. Better evidence for a warmer Martian climate comes instead from subtler water-worn features.

FACING PAGE: Floodwaters racing from southeast to northwest sculpted these teardrop-shaped islands in Ares Vallis. North is approximately up.

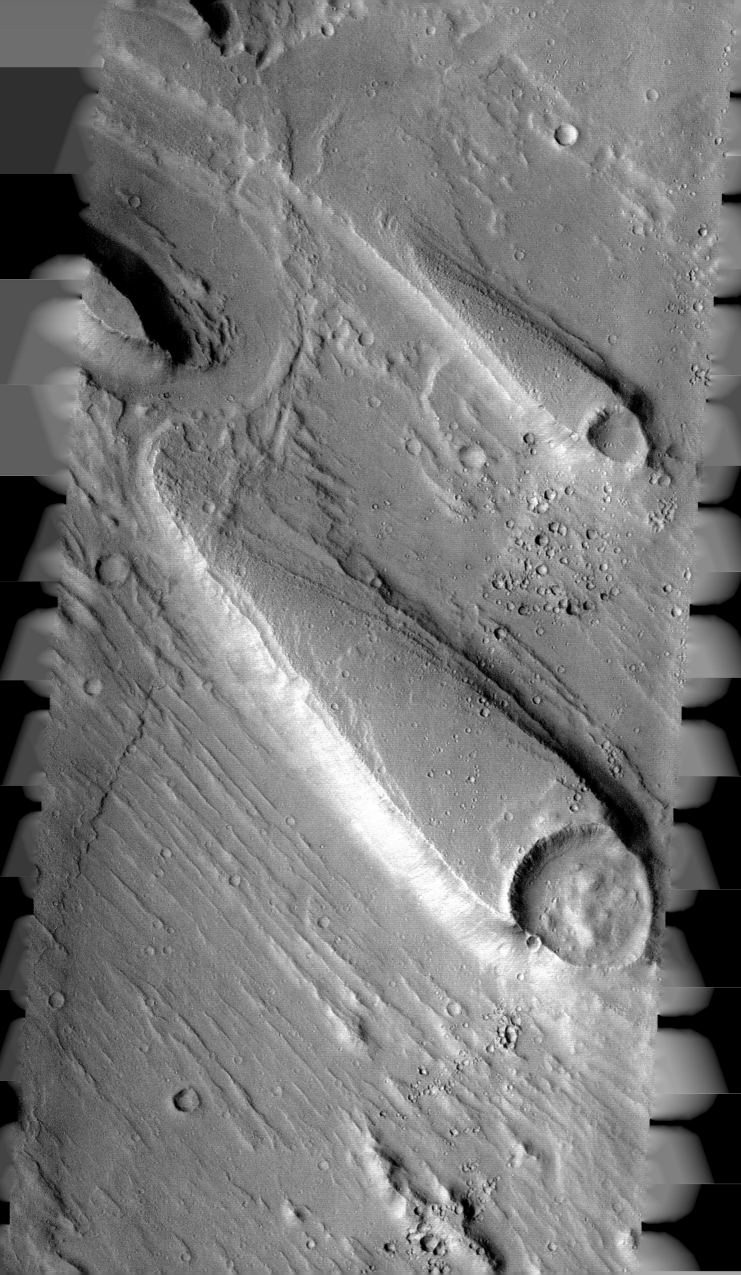

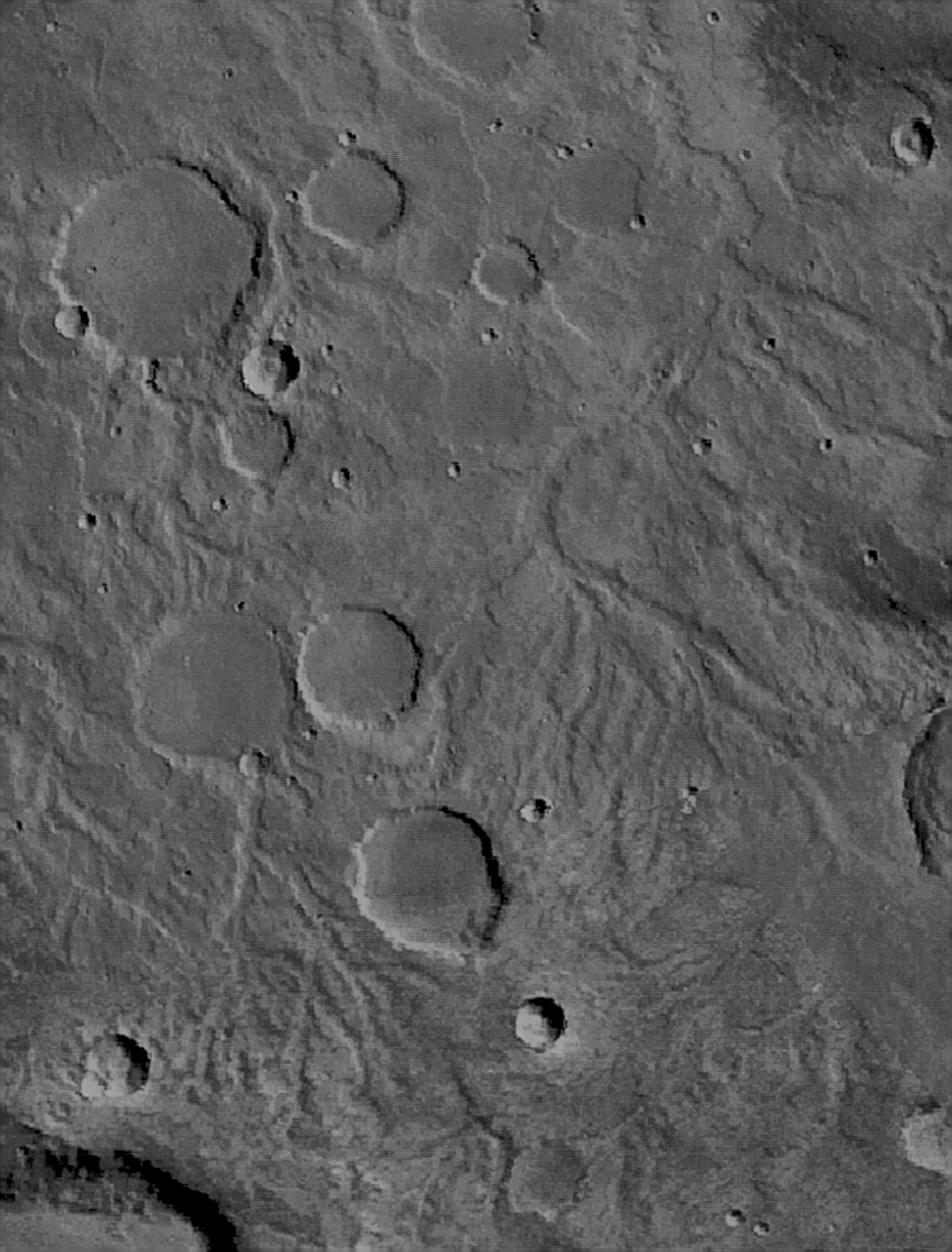

Rivers on Mars

RIVERS ONCE flowed on Mars. After previous spacecraft witnessed little but devastation, this was Mariner 9's greatest discovery. Unlike floods, rivers signify a prolonged flow of water, suggesting a warmer and wetter climate.

Mars has hundreds of dried-up riverbeds. Most cut through the cratered terrain that blankets the southern hemisphere, dissecting territory that dates back to the ancient Noachian period. The few exceptions appear primarily on steep slopes or on volcanoes, whose heat may have melted ice and created temporary rivers. Even on present Mars, a three-foot-deep river would take a week to freeze.

Like terrestrial rivers, and unlike the flood channels, most Martian rivers start small and grow larger downstream. However, Martian rivers are shorter than their earthly peers. Most are only 10 to 100 miles long; the longest runs 840 miles, about the length of the Rhine in Europe. By contrast, Earth's longest river, the Nile, is 4,160 miles long, nearly five times longer.

The rivers provide evidence that Noachian-era Mars had a higher erosion rate than today's Mars. Riverbeds that formed during much of the Noachian period are heavily eroded, whereas those which formed at its end look much fresher, indicating that the erosion rate plummeted when the Noachian period was yielding to the Hesperian, which in turn suggests that Mars had by then lost much of its air and water. From the eroded riverbeds as well as eroded craters and crater ejecta, scientists estimate that the Noachian erosion rate was roughly a thousand times

FACING PAGE: What look to be branching riverbeds billions of years old still exist on Mars. North is up.

the present rate. Still, that sounds better than it is. If it's correct, it means the Noachian erosion rate merely equaled that in terrestrial deserts.

Shower Power?

INDEED, Mars may never have seen rain. Proof of rain would be a key climatic clue, for rainfall demands an above-freezing temperature. Unfortunately, the Martian rivers don't demand rain. Antarctica—hardly a warm place—has a river, the Onyx, that flows only during the warmest days of summer. Some scientists have speculated that Martian rivers may have been equally miserly. On Earth most rivers collect water both from rainfall and from seepage of groundwater, a process called sapping. In contrast, the Martian rivers exhibit features suggesting they owed their power solely to sapping. For example, not only were the red planet's rivers shorter than Earth's, but they also had fewer branches and tributaries feeding them. Furthermore, those tributaries often begin as alcoves, another sign of groundwater sapping. Unlike rainfall, sapping does not require warm temperatures, because the rivers could have flowed beneath ice. On the positive side, however, even if groundwater sapping did power the rivers, the ground had to be rewetted. The simplest way to do that? Rainfall. Indeed, in 2002, Robert Craddock of the Smithsonian Institution and Alan Howard of the University of Virginia, in a paper titled "The Case for Rainfall on a Warm, Wet Early Mars," argued that ancient Mars must have had rainfall not only to rewet the ground but also to erode old craters.

In 1998 the Mars Global Surveyor scrutinized 500-mile-long Nanedi Vallis, which is northeast of Valles Marineris and takes its name from the word "star" in the South African language of Sesotho. Inside the valley the spacecraft's sharp eye spied a small channel, akin to the Colorado River inside the Grand Canyon. The Nanedi valley spans 1.5 miles, the channel just 660 feet. The spacecraft also spotted terraces in the valley, suggesting that during its life a river assumed different levels. Moreover, if Nanedi was carved by a river, the river meandered, typical of terrestrial rivers that have flowed awhile and indicating that this river flowed for a long time.

Unfortunately, this interpretation has not worn so well with time. For one thing, Mars Global Surveyor has failed to find a channel anywhere else in Nanedi Vallis. Also, Nanedi itself has few tributaries, indicating again that the river may have been fed from below, by sapping of groundwater, rather than from above, by rainfall. Nevertheless, if Nanedi was once a river, it may have been one of the last to flow on Mars, since the valley is of Hesperian age and appears quite fresh—little erosion has occurred since the river dried up.

Surprisingly, water may have flowed much more recently. In 2000, Michael Malin and Kenneth Edgett of Malin Space Science Systems in San Diego

presented evidence for flows of water within just the past few million years. In images from the Mars Global Surveyor, they identified thousands of gullies that resemble gullies in Canada, Greenland, and Iceland dug by water. The Martian gullies appear beneath the brinks of steep slopes, possess channels, and spill into debris aprons. They are young: most gullies bear no craters at all. Over 90 percent are in the southern hemisphere, and all are poleward of 27 degrees latitude. Many of the gullies occur in craters; many others are in two valleys, Dao Vallis and Nirgal Vallis, near the southern impact basins Hellas and Argyre, respectively. Strangely, most of the gullies appear on slopes facing the pole and thus away from the Sun. Perhaps subsurface ice traps water that bursts forth during warm spells. In 2003, Philip Christensen of the University of Arizona suggested instead that the gullies form when snow melts. In his view, periods of high axial tilt cause the poles to warm and snow to migrate from the poles to pole-facing slopes at mid-latitudes. Then, during periods of low axial tilt, the mid-latitudes warm, and the snow slowly melts, creating the gullies. However, some scientists believe water did not produce the gullies. Instead, they say, the gullies were carved by outflows of carbon dioxide.

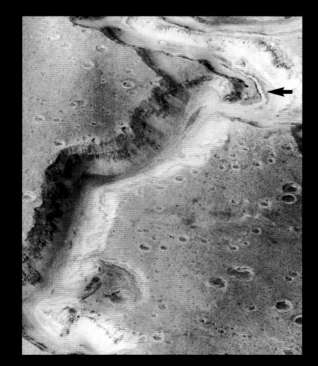

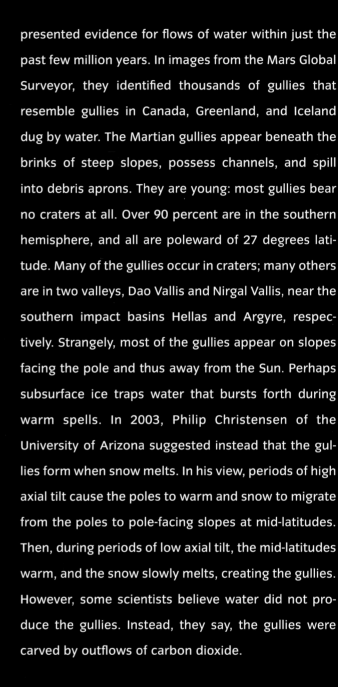

LEFT AND FACING PAGE: This close-up from Mars Global Surveyor reveals a possible ancient riverbed (arrow) within Nanedi Vallis. North is up.

FOLLOWING PAGES:

Page 130: Water gushing out of this crater wall probably carved these gullies. The crater, named Newton, resides in Terra Sirenum.

Page 131: Recent flows of liquid water likely cut the gullies in this crater wall, part of Noachis Terra. The image is about two miles across. North is up.

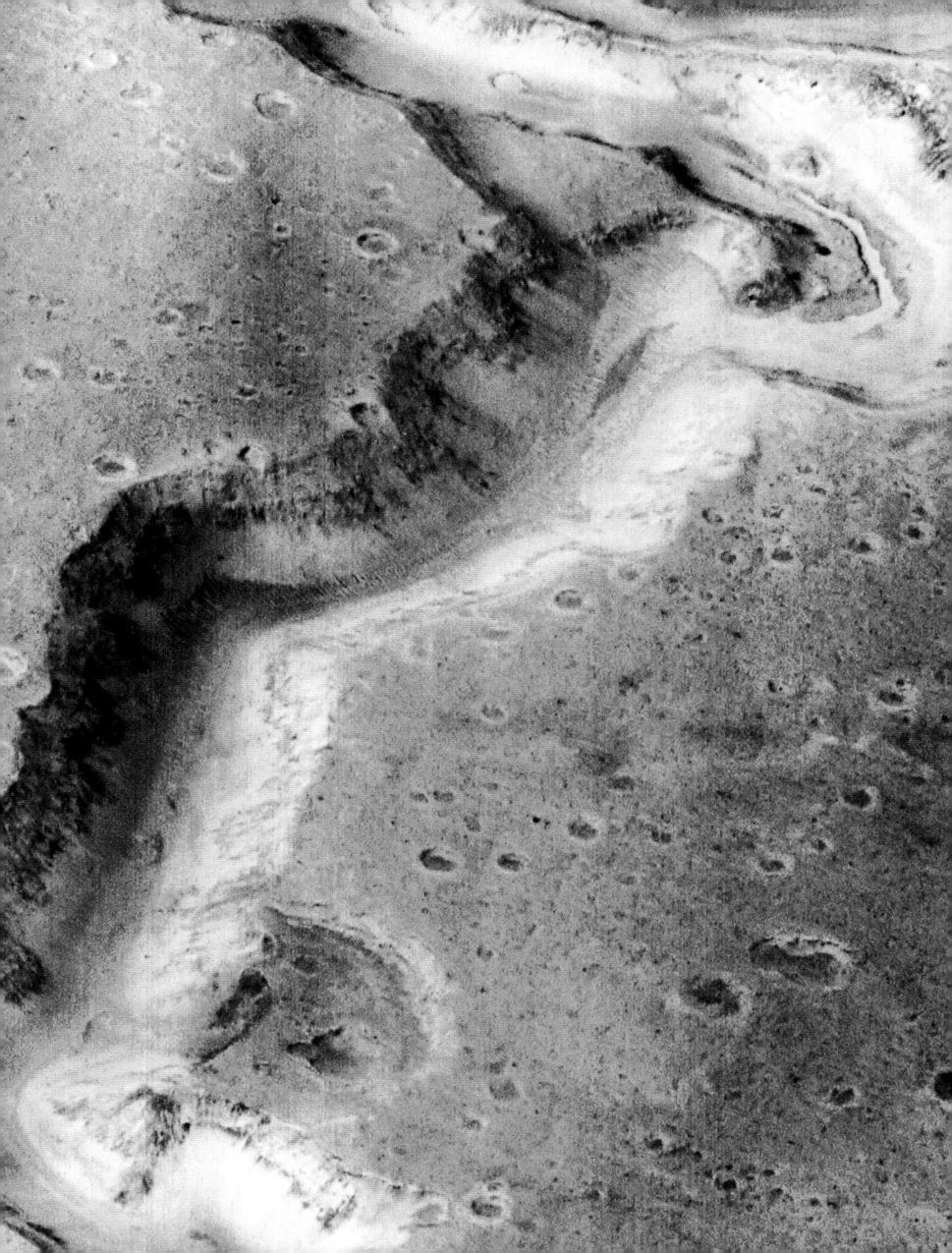

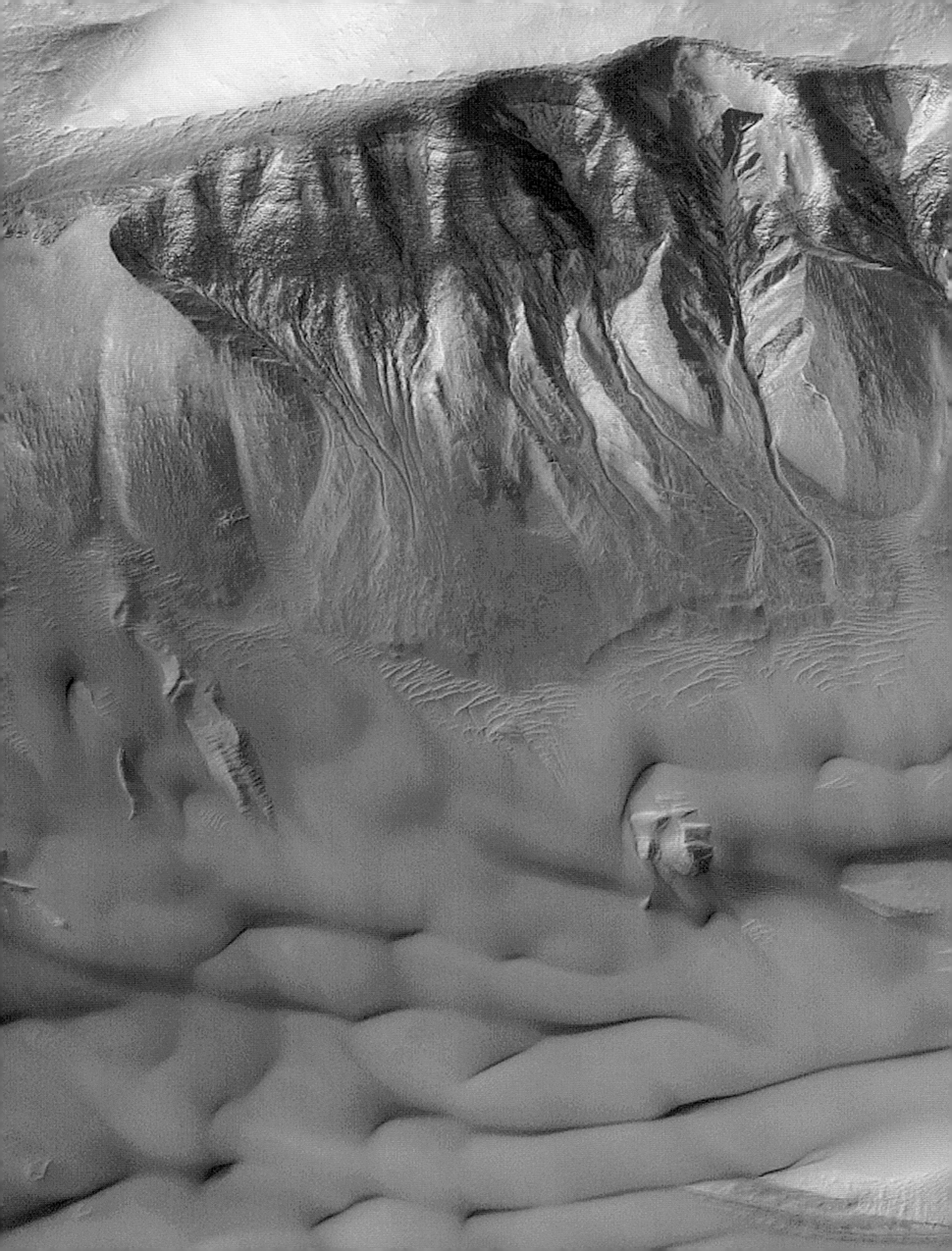

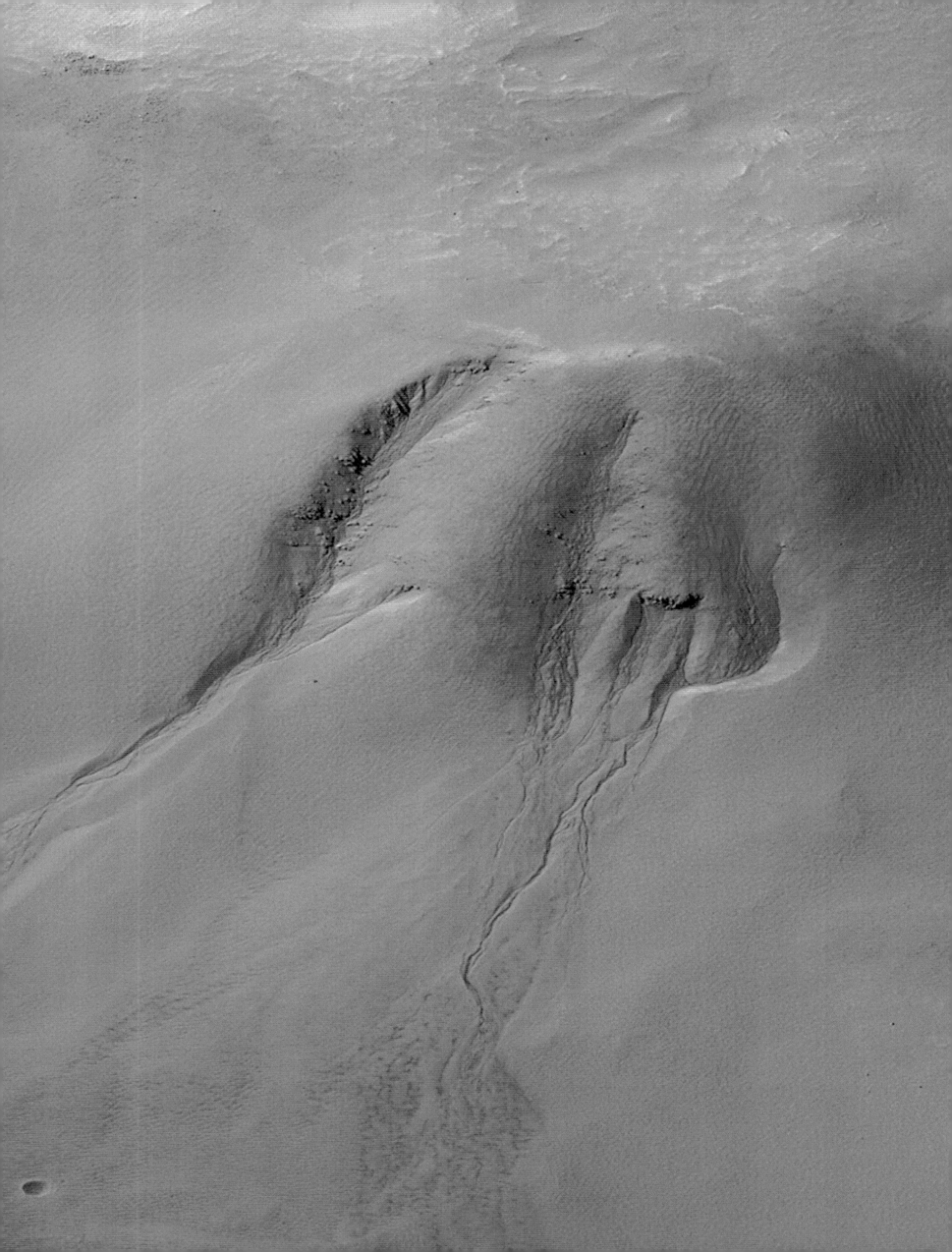

Lakes on Mars

MORE THAN just rivers may have danced over the Martian surface. Lakes may have dotted the planet, too. The same craters that rendered the first spacecraft portraits so bleak could have filled with water, whose waves lapped shores glistening in sunlight.

The largest and deepest potential lake was the southern impact basin Hellas, which formed when Mars was young, during the Early Noachian period. In 2001, Jeffrey Moore of NASA's Ames Research Center and Don Wilhelms of the U.S. Geological Survey marshaled a variety of evidence to propose that Hellas had once been a lake, probably covered with ice. First, they noted, two channels east of Hellas—Dao Vallis and Harmakhis Vallis—look as though they carried water into a lake. On the topographic map these channels appear purple, denoting a low elevation, and cut through green, turquoise, and blue terrain. Second, "bathtub-ring" deposits surround Hellas, suggesting the lake achieved different heights at different times. The most pronounced contour is 5.8 kilometers below Martian sea level, corresponding to purple on the topographic map, and a lesser contour has a depth of 3.1 kilometers, green on the map. Third, both Dao Vallis and Harmakhis Vallis are prominent above the 5.8-kilometer contour but look softer below it, indicating the rivers discharged water into a lake of this depth. Fourth, the Hellas floor is fine-grained, suggesting deposits from water. Fifth, parts of western and northwestern Hellas resemble a honeycomb, which may have formed when the lake began to lose water through evaporation off its icy surface, causing ice blocks atop the lake to hit the soft muddy bottom. If Hellas was indeed a lake, it could have dwarfed any on present Earth. In fact, a lake filling the entire basin would span 1,400 miles, the diameter of Pluto and the distance from Pittsburgh to Denver. Furthermore,

because of its great depth, a fully filled Hellas would have held nearly as much water as all the northern plains put together. The other large impact basin in the southern hemisphere, Argyre, may also have hosted a lake, albeit one shallower than the Hellas lake. A channel named Uzboi Vallis leads out of Argyre to the north.

Smaller craters may have held smaller lakes. Nathalie Cabrol of NASA's Ames Research Center and her colleagues have called attention to the 105-mile-diameter crater Gale, south of the Elysium volcanoes, which squats on the boundary between the lowlands and the cratered highlands. This crater displays streamlined terraces, which may be former shorelines; also, channels north of Gale feed into it. In addition, sedimentary deposits on the crater's floor have a crescent shape, like a remnant of milk in a tilted glass. These crescent-shaped deposits may have formed when the lake dried up. To the east, another crater on the lowland-highland boundary, Gusev, sits south of the isolated volcano Apollinaris Patera. A long riverbed, Ma'adim Vallis, flows into Gusev from the cratered highlands. Intriguingly, Cabrol's team says that both Gale and Gusev have young floors, so their lakes may have existed recently, during the Amazonian period. Perhaps the nearby volcanoes melted ice and caused water to flow into the craters.

In 2000, Michael Malin and Kenneth Edgett presented additional evidence for former Martian lakes when they unveiled images from the Mars Global Surveyor that they said showed sedimentary rocks. On Earth such rocks can form when dirt falls to the bottom of a lake and gets compressed. Malin and

Edgett identified the sedimentary rock layers in several areas of Mars, including Valles Marineris, implying that the canyons once held lakes. Beneath the enormous cliffs the spacecraft's sharp-eyed camera spotted no boulders, suggesting to Malin and Edgett that the canyon walls consist of just the fine-grained materials which a lake should precipitate: clay, silt, sand. However, some scientists think the layers are instead layers of volcanic ash that had nothing to do with water. A geologist who picked up the rocks could quickly tell the difference. If the rocks are indeed sedimentary, they may hold fossils of Martian life.

FOLLOWING PAGES:

Pages 134-135: On ancient Mars the Hellas impact basin may have been a large lake. Here Hellas appears as a light-colored oval surrounded by cratered highlands. Notice the valleys on the east side, feeding into Hellas in a northeast-to-southwest direction. North is up.

Page 136: This topographic sphere is centered on Hellas, the deep purple scar. The impact that created Hellas excavated so much rock that it colored much of the rest of this hemisphere topographic orange and red. Again notice the valleys, purple here, feeding into eastern Hellas. North is approximately up.

Page 137: This close-up shows the valleys in eastern Hellas that may have carried water into the basin. From north to south, the three valleys are Dao Vallis; Niger Vallis, which joins Dao; and Harmakhis Vallis. North is up.

Page 138: The Argyre impact basin may have hosted a lake smaller than the one in Hellas. North is up.

Page 139: These rock layers on the floor of Valles Marineris—specifically, southwestern Candor Chasma—may have been laid down by lakes in the canyon. The image here is not quite a mile across, and each layer is about ten yards thick. North is up.

Pages 140-141: A river likely flowed into Gusev Crater, creating a possible lake there. The crater is 100 miles across; the riverbed, Ma'adim Vallis, is 370 miles long. North is to the left.

Tales from Topographic Oceans

IF MARS HAD LAKES, why not an entire ocean? On the topographic map the northern lowlands bear the suggestive color blue, and this indeed is the land any Martian ocean would have occupied: three fourths of Mars drains into the northern plains. The large flood channels that empty into Chryse may have contributed their water to this ocean. Indeed, those floods may have created the ocean.

A Martian ocean was first suggested in the 1980s, when scientists such as the Jet Propulsion Laboratory's Timothy Parker used Viking Orbiter images to identify possible shorelines on the northern plains. In the early 1990s, Victor Baker of the University of Arizona and his colleagues offered a scenario for the ocean's formation: the Tharsis volcanoes erupt, melting ice and causing floodwaters to surge northward. Because the volcanoes emit carbon dioxide, water vapor, and sulfur dioxide, these greenhouse gases warm Mars and melt more ice, augmenting the ocean. When the volcanoes shut down and the air thins, Mars cools, the ocean freezes, and the ice sinks into the ground. There the ice stays, until the volcanoes erupt again.

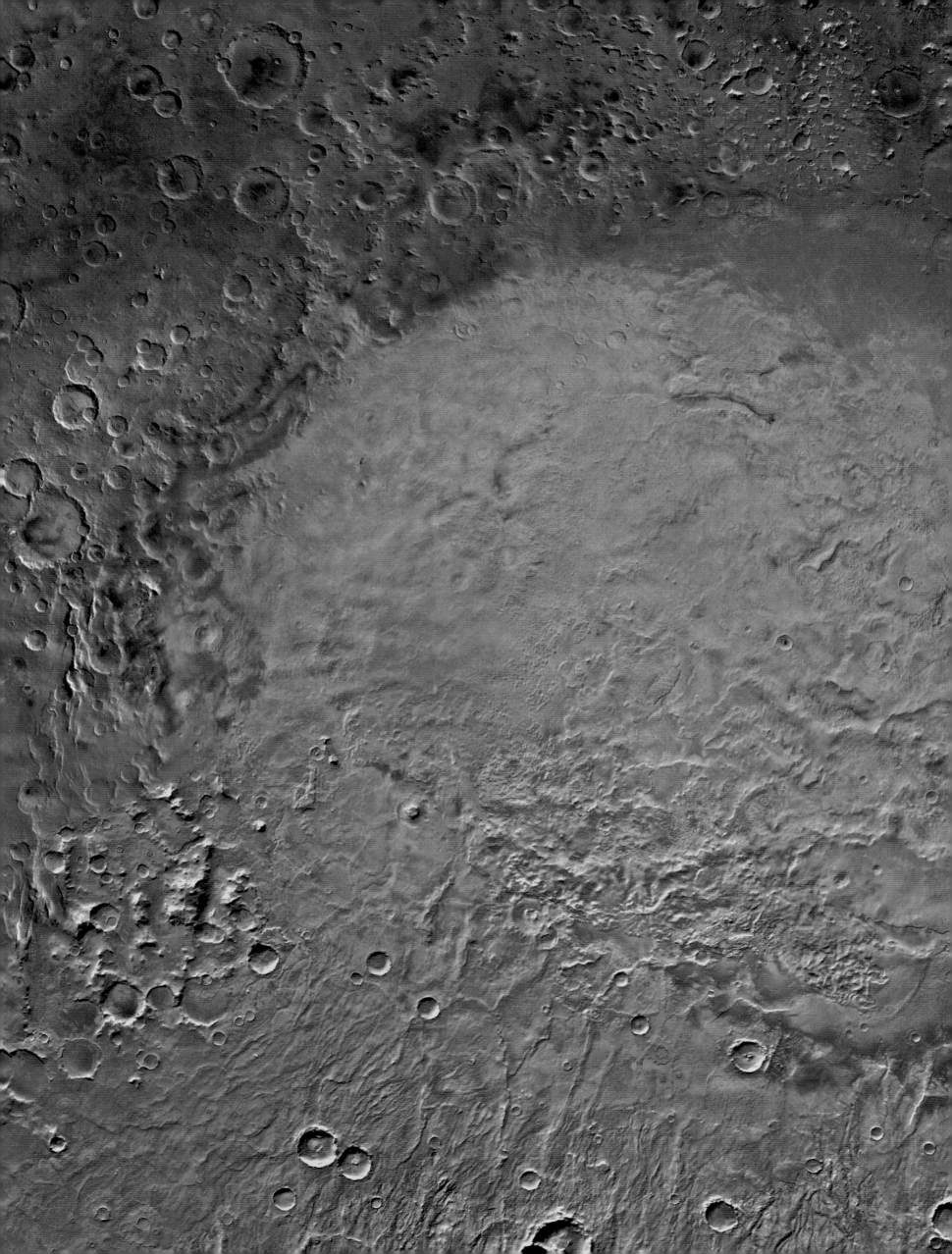

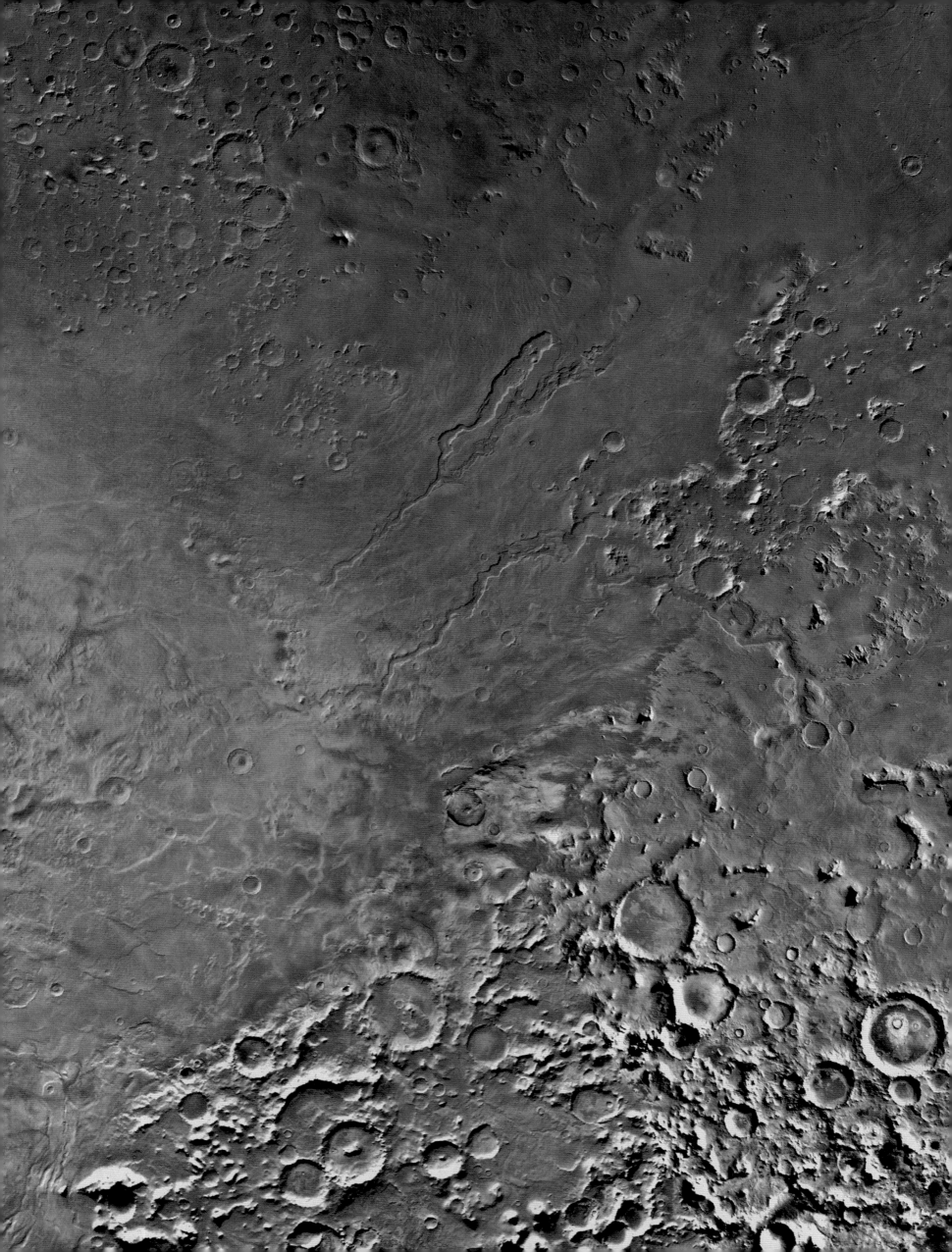

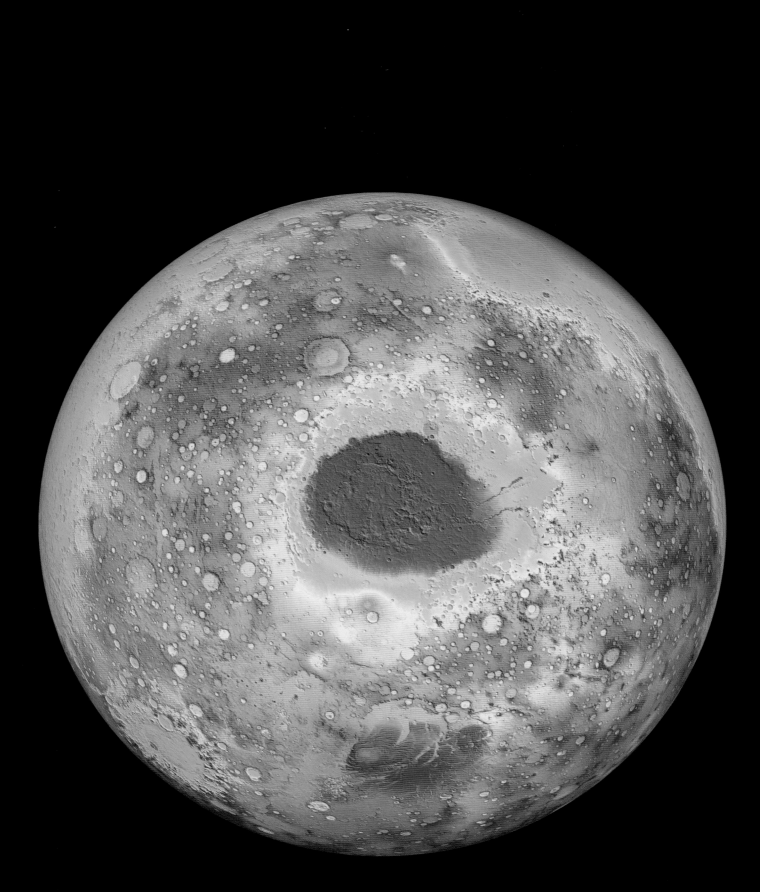

ELEVATION

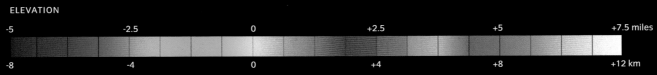

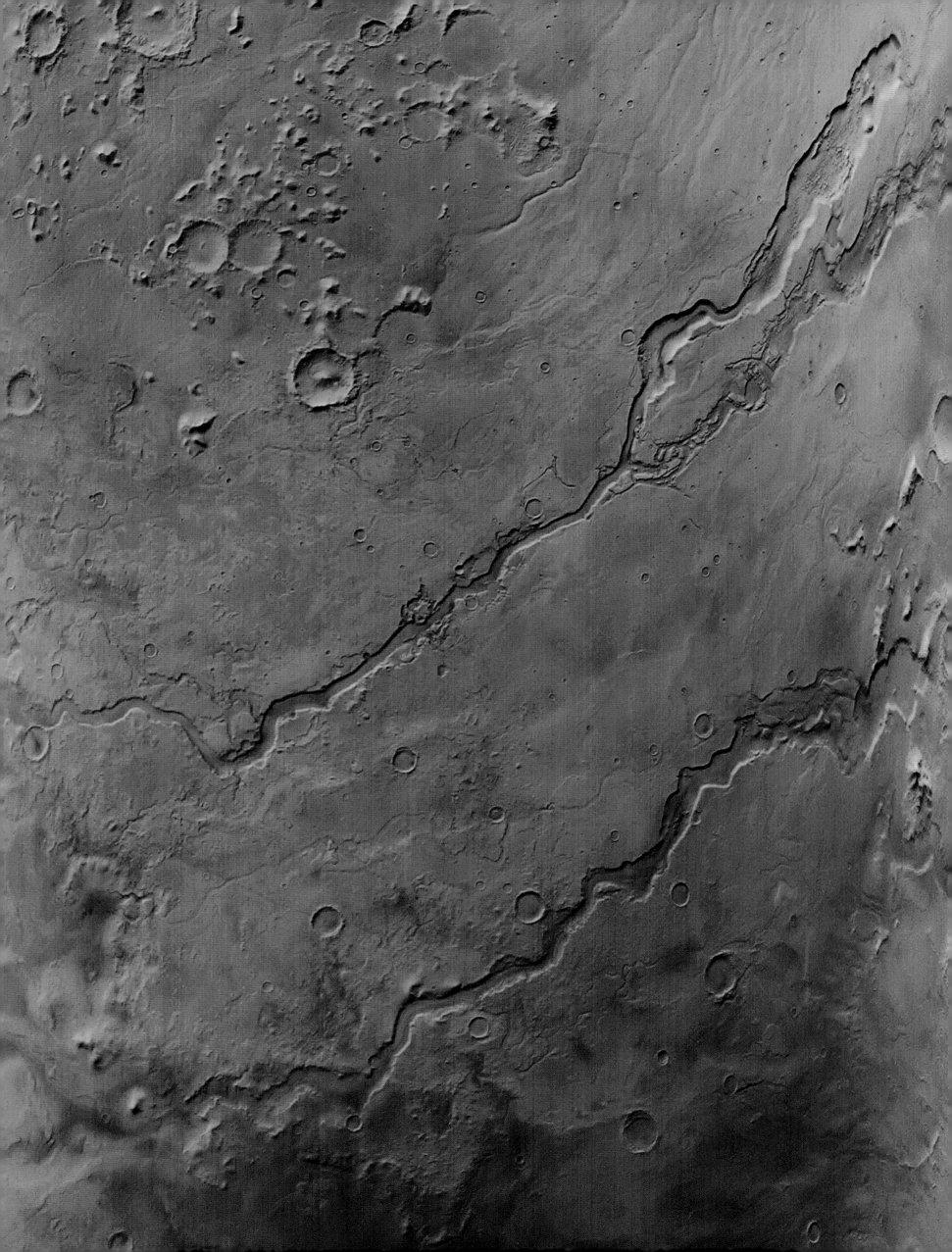

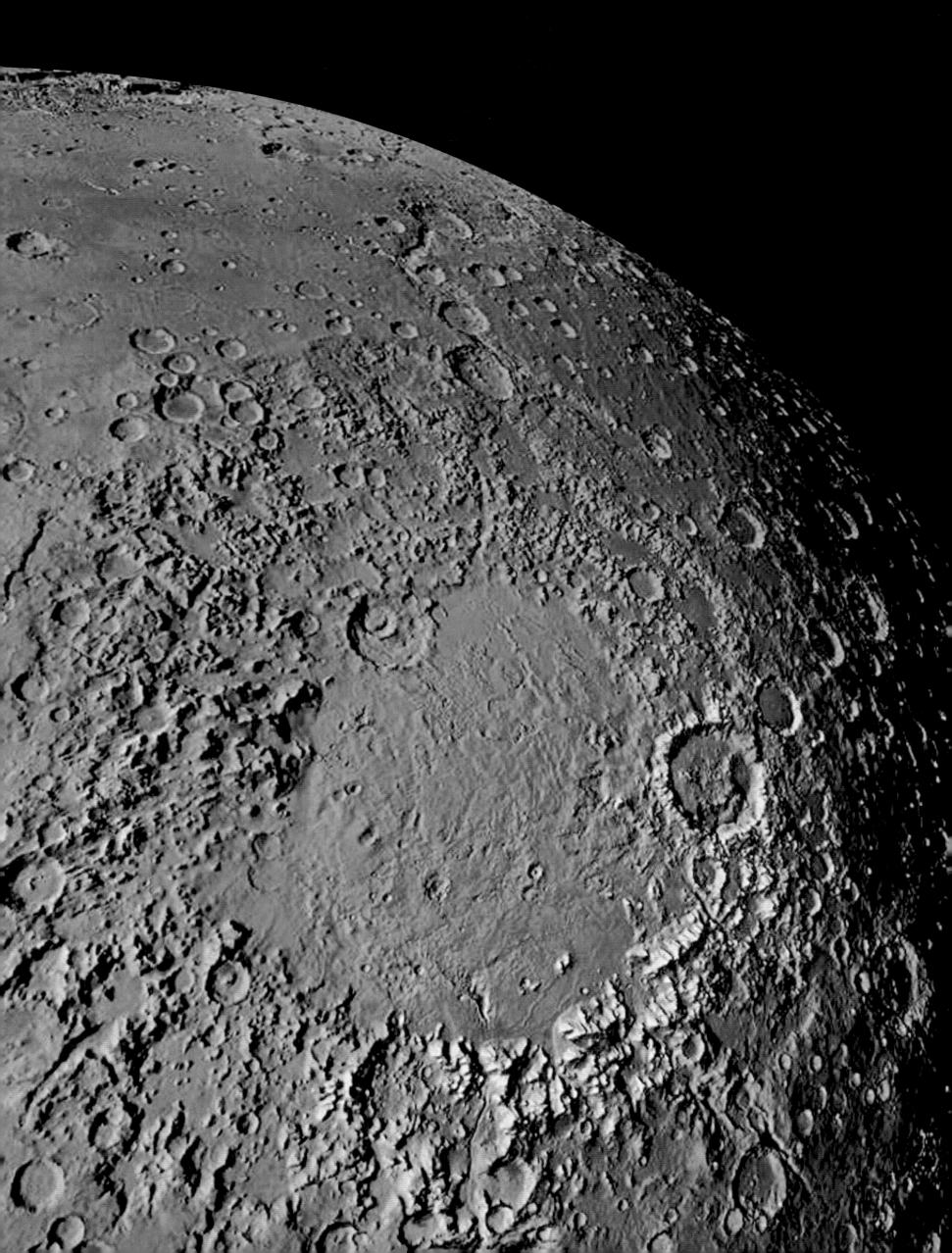

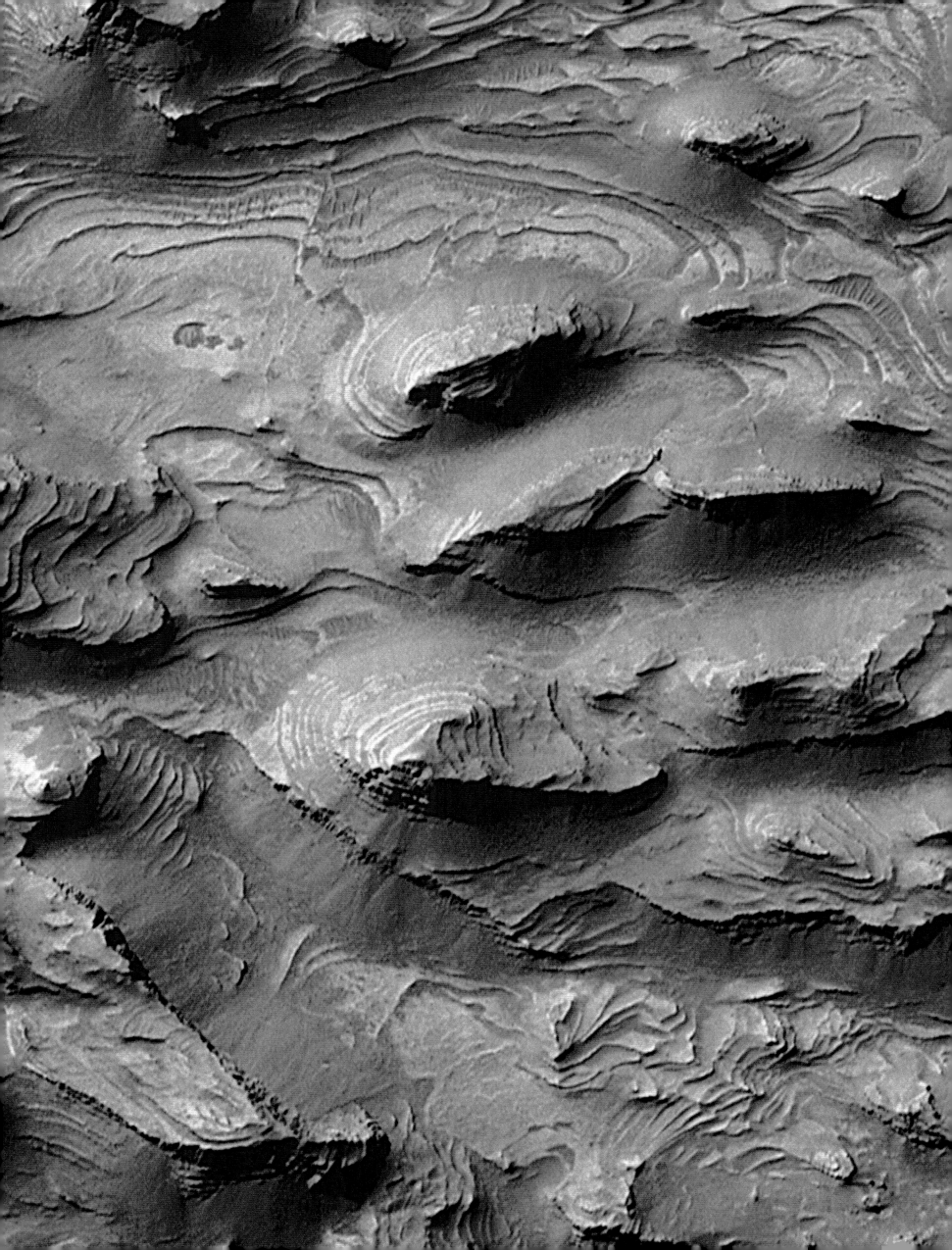

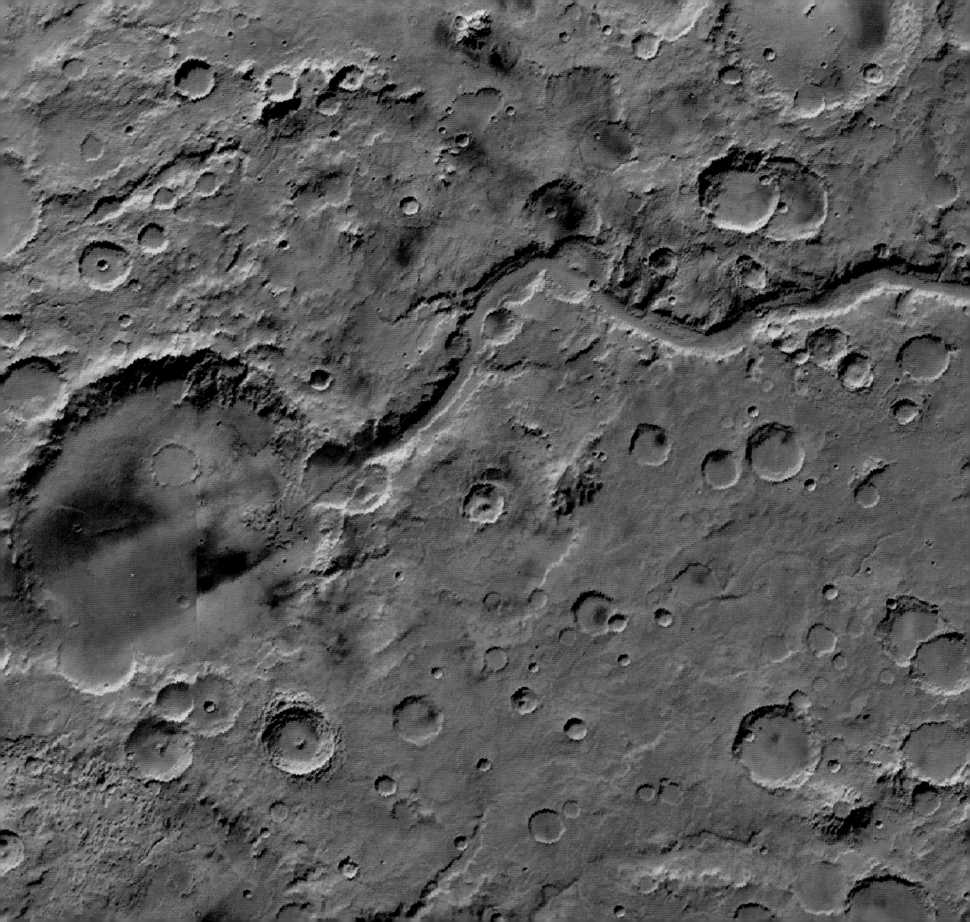

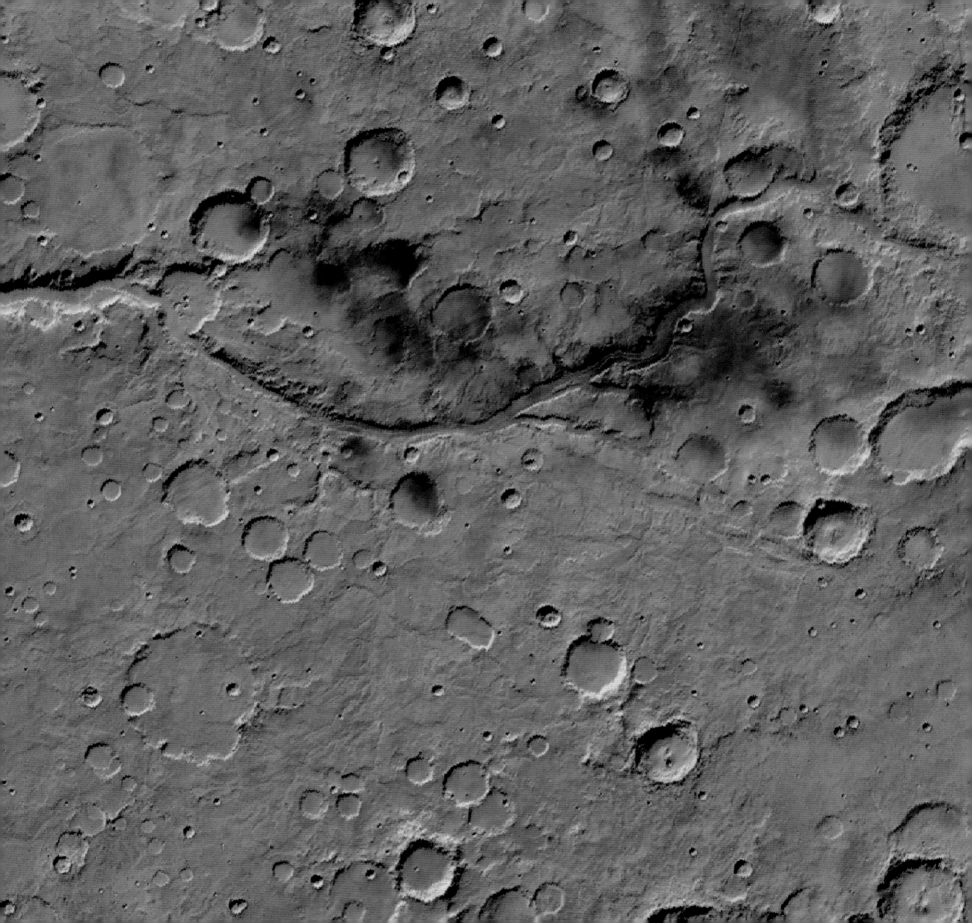

The Martian surface
is smoother inside this
proposed shoreline
than outside, another
encouraging sign

than a few hundred meters. The three exceptions are a crater named Lyot, presumably because the impact disturbed the land; and the volcanic provinces of Tharsis and Elysium. Furthermore, the Martian surface is smoother inside this proposed shoreline than outside, another encouraging sign. And terraces exist near the shoreline, which may have formed when the water receded. So the ocean hypothesis may hold water after all. However, the Mars Global Surveyor's camera imaged the proposed shorelines and saw none, leading critics to proclaim the ocean idea all wet.

However, the notion of an ocean has caused commotion. Many scientists support it, many oppose it. The Mars Global Surveyor supplied ammunition to both sides. On the positive side, the spacecraft's laser altimeter revealed that the northern lowlands are exceedingly flat—as if they once held an ocean, because not only would the water press down on the land, but it would also carry dirt to fill in small depressions. On the topographic map, apart from a few craters, the northern plains look featureless, a nearly uniform sea of blue, with no great hills or valleys.

On the negative side, the Mars Global Surveyor's laser altimeter showed that one proposed ocean shoreline weaves up and down by over 6 miles—bad news for the ocean hypothesis, since a shoreline, like the ocean itself, should have had the same height everywhere. However, the shoreline of a smaller proposed ocean, containing one twentieth of the volume of water in the Atlantic Ocean, does have a more uniform altitude. With three exceptions, the altitude of this inner shoreline never deviates by more

FACING PAGE: This and the following three topographic maps show the northern hemisphere of Mars as water (black) fills it. The north polar cap is at center; the red and white peak at one o'clock is the volcano Elysium Mons; the red patch at nine and ten o'clock is the Tharsis bulge; and the white peak between ten and eleven o'clock is the volcano Olympus Mons.

The facing page shows the northern lowlands flooded to a depth of 500 meters. The black patch at two o'clock fills part of the Utopia impact basin, near the Viking 2 lander; a slight rise separates this water from water closer to the north pole.

FOLLOWING PAGES:

Page 144: When water reaches a depth of 1,000 meters, the two water bodies connect.

Page 145: Flooded to a depth of 1,490 meters, the ocean matches the proposed shoreline that Mars Global Surveyor's laser altimeter found had a fairly uniform elevation.

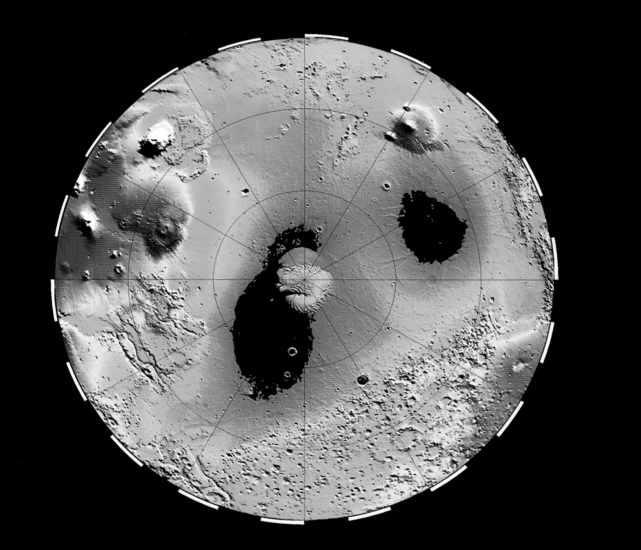

ELEVATION

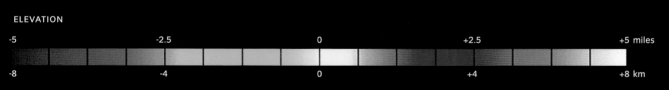

-5 -2.5 0 +2.5 +5 miles

-8 -4 0 +4 +8 km

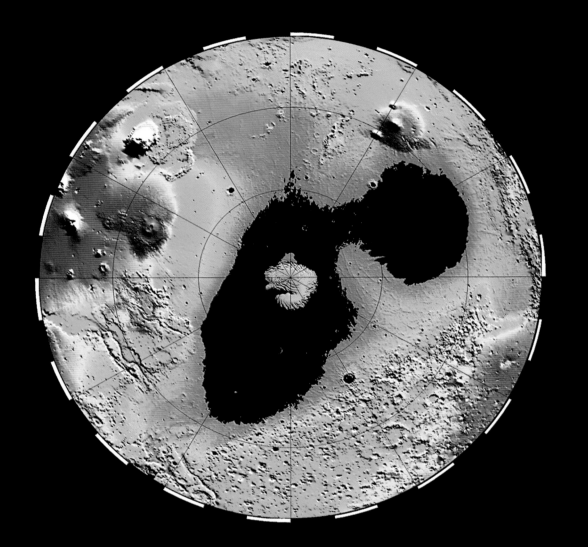

ELEVATION

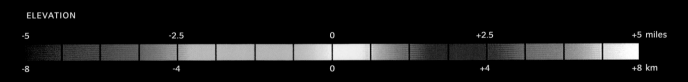

-5		-2.5		0		+2.5		+5 miles
-8		-4		0		+4		+8 km

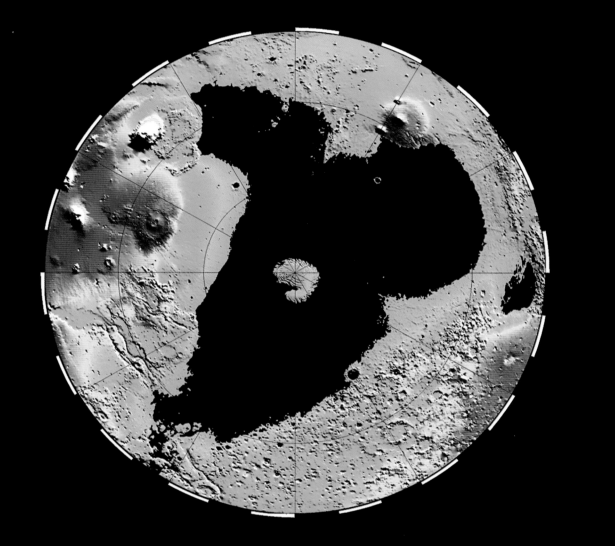

-5   -2.5   0   +2.5   +5 miles

-8   -4   0   +4   +8 km

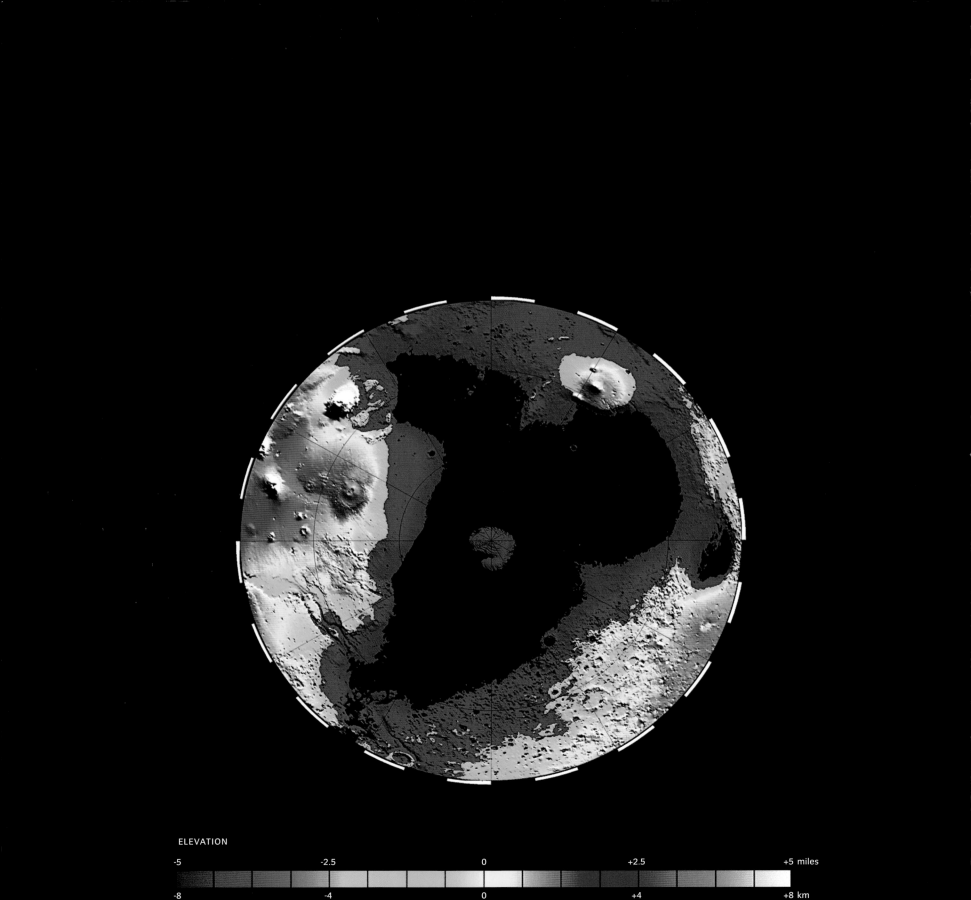

ELEVATION

| -5 | | -2.5 | | 0 | | +2.5 | | +5 miles |

| -8 | | -4 | | 0 | | +4 | | +8 km |

Clues from Surface Composition

IF WATER once occupied parts of the Martian surface, it should have altered the land's chemical and mineral composition. At present, however, clues from the surface composition point different ways.

In 1998 the Mars Global Surveyor's Thermal Emission Spectrometer favored the pro-water side when it discovered large quantities of hematite. Hematite (Fe_2O_3) is the most common iron ore on Earth. It ranges in color from reddish brown to gray to black, and when it scratches a piece of unglazed porcelain it produces a red streak. Most of the Martian hematite occupies an equatorial area 300 miles across, in Meridiani Planum, east of Valles Marineris and southeast of the flood channels that spill into Chryse Planitia. The Meridiani region gets its name because it's where 0 degrees longitude passes—the Martian equivalent of Greenwich, England. Back in the nineteenth century, mapmakers Johann Mädler and Wilhelm Beer chose a dark round patch here as a reference for measuring the planet's rotation period. The hematite in Meridiani Planum has coarse grains, just the type that iron-rich water would precipitate. The water may have been hot, since heat speeds up chemical reactions. If so, the water may have been a hot spring, providing not only liquid water but also a source of energy—two things life seems to need. Hematite also occurs in a 170-mile-wide equatorial crater west of Meridiani Planum named Aram Chaos. It's near Ares Vallis, the flood channel upstream from the Pathfinder spacecraft.

The surprisingly high levels of chlorine and sulfur that the Viking landers found in Martian soil may also support the idea of past water, especially a northern ocean, since both spacecraft landed in the northern plains. As described in EARTH, the chlorine may be in salt (NaCl) that blanketed the land after the water dried up, and the sulfur may be in sulfates that precipitated out of the ocean. On the other hand, the two elements may merely be volcanic deposits such as hydrochloric acid (HCl) and sulfur dioxide (SO_2). Furthermore, the Mars Global Surveyor's Thermal Emission Spectrometer has seen no sulfates at all.

In 2000 the Mars Global Surveyor's Thermal Emission Spectrometer detected a mineral suggesting Mars has long been dry. Olivine is a greenish iron-magnesium silicate that as a gem is called peridot. It makes up much of the terrestrial mantle but rarely crops up on the surface, because it quickly succumbs to a wet atmosphere. Hawaii's "green beaches," erupted by volcanoes, are one of the few terrestrial examples. The olivine on Mars indicates that the red planet hasn't experienced much of a warm, wet atmosphere during the past several billion years.

Worse, the same instrument that detected hematite and olivine has failed to find carbonate rocks like limestone. On Earth such rocks form because of water. Rain removes carbon dioxide from the air, creating carbonic acid. This acid wears away rocks, carrying their calcium and magnesium into lakes and oceans, where organisms use them to construct shells

FACING PAGE: Further flooding attempts to match a larger proposed ocean, but the originally suggested shoreline

that eventually become carbonate rocks. So rapid is this process that if volcanoes stopped replenishing atmospheric carbon dioxide, it would vanish from the air just 10,000 years later, killing all plant life. Even if Mars never had rain or life, it should have carbonate rocks—*if* it had large bodies of water. That's because even without rain, carbon dioxide dissolves in water, and even without life, that dissolved carbon dioxide precipitates to the seafloor and becomes carbonate rocks. If Mars had an ocean, a carbon dioxide atmosphere thicker than Earth's should have precipitated nearly 100 meters of carbonate rocks. Yet they haven't been found. The most straightforward explanation is that the Martian surface never had much water. However, groundwater may have carried the carbonates underground, where the spacecraft couldn't see them. Also, ultraviolet light may have destroyed surface carbonates. Or the ocean may have been ice-covered and thus out of contact with the carbon dioxide atmosphere.

Fortunately, scientists don't have to study the Martian surface from afar; they can also study the Martian meteorites. Some carbonates exist in the old Martian meteorite ALH 84001. A younger Martian meteorite, named Lafayette, bears iddingsite, a mix of rust and clay that forms in water. The iddingsite contains the noble gases krypton and xenon, captured from the Martian atmosphere, which means the rock sat on or near the red planet's surface. Together with Lafayette's young age—1.3 billion years—this indicates that water may have flowed on the Martian surface recently. The composition of another Martian meteorite, named Shergotty, reveals that the magma the volcanoes erupted contained water. According to a 2001 study, pre-eruptive magma was as much as 1.8 percent water, so the volcanic eruptions must have released large quantities of water and water vapor, which warmed Mars through the greenhouse effect.

Ice on Mars

MARS IS certainly not warm today. As a result, nearly all of its water is frozen. The Viking 2 lander, at latitude 48 degrees north, saw some of this ice for itself. During winter, when the temperature fell to the frost point of carbon dioxide (about −195 degrees F.), frost covered the ground; however, even when the temperature rose above the carbon dioxide frost point, the frost remained. Thus, the initial frost probably consisted of both carbon dioxide and water, but when the surface warmed, the carbon dioxide vaporized, leaving the water ice behind. Toward spring, as the ground warmed further, this also vanished.

The most obvious Martian ice huddles in the polar caps, which wax and wane with the seasons. At the start of northern fall, the north polar cap is small, extending down to a latitude of about 80 degrees north. During fall, clouds called the polar hood form over and beyond the polar cap and soon extend down to a latitude of 50 degrees. Under the cover of the polar hood and the darkness of winter, the ice cap expands. By the start of spring, the polar hood begins to clear, and observers see the northern cap at its greatest, stretching down to about 65 degrees latitude. As spring advances, however, the ice begins to warm and retreat. During midspring the cap retreats about 10 miles a day, shrinking to its minimum size shortly after the start of summer.

FACING PAGE: Frost comes to Utopia, the Viking 2 landing site.

Similar events unfold in the south. However, instead of a polar hood, separate clouds appear. Also, because southern winter is longer and colder, the south polar cap grows larger, reaching as far as 50 degrees south latitude. It even extends into the southern impact basins Hellas and Argyre, whose floors appear bright during the winter. When, at winter's end, the ice begins to retreat, it does so more irregularly, because of the rougher topography. Also, it vaporizes more slowly, because it is brighter: during spring the northern cap reflects about 64 percent of the sunlight hitting it, the southern cap about 82 percent. That's probably because the northern ice cap forms during northern winter, or southern summer, when dust storms darken its ice.

The polar caps harbor water ice

As briefly described in EARTH, each polar cap is really two caps in one. The more dynamic and superficial is the seasonal cap, which comes and goes with the seasons and is mostly carbon dioxide. The other cap, the residual cap, survives the summer heat and is mostly water ice. As mentioned, the seasonal cap gets larger in the south than the north: during northern winter, when the north polar cap is largest, 10 to 15 percent of the Martian atmosphere disappears; but when the south polar cap peaks, during southern winter, 25 percent of the atmosphere vanishes. The *residual* cap is just the opposite, however: larger in the north than the south—600 miles across versus 200 miles.

According to the Mars Global Surveyor's laser altimeter, the north polar cap sits in a basin, the south polar cap on a rise. The altimeter was so precise that it even saw the changing seasons cause the caps to rise and fall a few feet. Both residual caps look like spirals. Earth's poles have nothing similar. In the north, a large canyon named Chasma Boreale runs away from the polar cap for 400 miles. Some scientists have suggested it might be a flood channel, formed during a climate cycle that warmed the poles, but Mars Global Surveyor failed to find any flood features, such as streamlined islands.

The polar caps harbor water ice. To quantify a planet's water supply, scientists calculate the depth the water would have if they spread it evenly over the planet. All of Earth's water—liquid, frozen, and gaseous—would yield a global ocean 1.7 miles, or 2.7 kilometers, deep. Martian numbers are more modest. As discussed in AIR, the atmosphere is so dry that its water would amount to an "ocean" only 10 microns deep. Fortunately, the polar caps contain a good deal more. In 1999 the Mars Global Surveyor's laser altimeter indicated that the smaller southern residual cap actually has about twice as much water ice as its northern counterpart. The southern residual cap has as much ice as Greenland. Combined, the north and south residual caps possess an eighth as much water ice as Antarctica. If it all melted to form a global Martian ocean, the water would be only 22 to 33 meters deep— probably not enough to explain the ancient rivers and floods, let alone lakes and the possible ocean. For example, Michael Carr of the U.S. Geological Survey has estimated that the flood channels and riverbeds demand at least several hundred meters of water— over ten times what's in the polar caps.

FACING PAGE: Poles apart: The north polar cap of Mars. The dark belt around the ice consists of sand dunes.

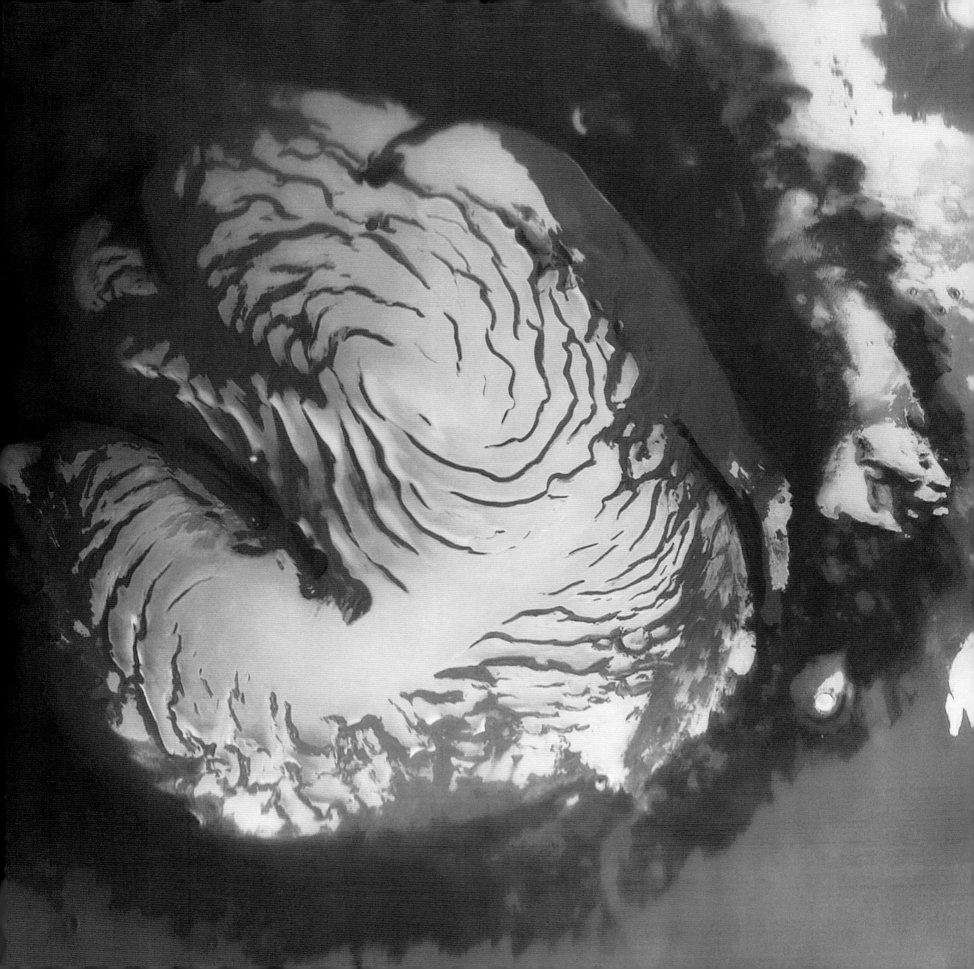

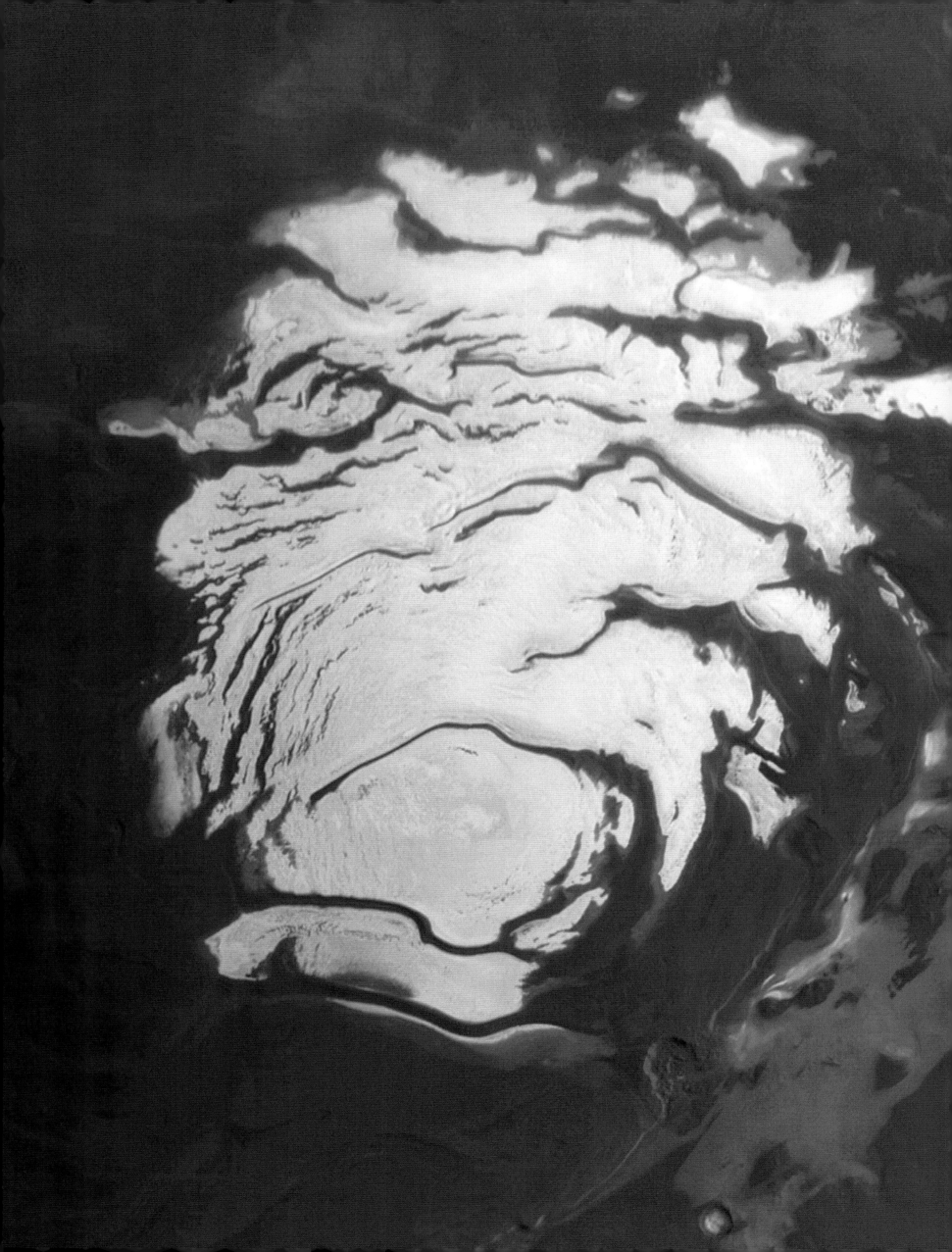

So where did it go? With only a pittance at the poles, much of it likely went underground and froze. Scientists had long expected subsurface ice poleward of 40 degrees latitude. At these high latitudes, subsurface temperatures below −105 degrees F. persist year-round, so water ice survives—it doesn't vaporize. In contrast, nearer the equator, ice can exist only deeper beneath the surface, because summer warmth vaporizes any ice near the surface.

Martian craters suggest just such a pattern: shallow ice at high latitudes, deeper ice at equatorial ones. Aptly named "splosh" craters are key indicators of subsurface ice. They look like what would happen when an asteroid hits mud, shooting lobes of wet material away from the resulting crater. Splosh craters exist at all Martian latitudes. Near the equator, however, only craters larger than 3 or 4 miles exhibit the splosh feature; there the ice is buried so deeply that only the largest impactors disturb it. By contrast, at latitudes of 50 to 60 degrees, craters just a mile wide show the splosh, indicating ice much closer to the surface. In addition, craters poleward of 30 degrees look softer, probably because the craters rest on icy ground that over time partially crumbles.

Further evidence for Martian ice comes from Martian fire. Volcanoes can melt ice, and west of the volcano Elysium Mons, channels that probably carried water extend for hundreds of miles, complete with teardrop-shaped islands. Also, the channels leading into the Hellas impact basin may owe their existence to the volcanoes on its rim. These channels indirectly suggest that the much greater Tharsis volcanism melted sufficient ice to create a northern ocean.

FACING PAGE: The south polar cap of Mars.

FOLLOWING PAGES: Deep blue designates regions of Mars that the Odyssey spacecraft found rich in hydrogen and thus presumably water ice. When this map was made, carbon dioxide frost cloaked the north pole, hiding water ice there. North is up; zero longitude is at center.

Furthermore, the Tharsis lava itself contained enough water to form a global ocean 120 meters deep. Still more evidence for volcanically related water comes from signs of explosive eruptions. As mentioned in FIRE, if water invades their vents, even gentle volcanoes can erupt violently, producing ash—and ash deposits may lie near Elysium Mons and the Hellas volcano Tyrrhena Patera. Ash deposits may also surround Alba Patera—the great but peculiar Tharsis volcano far to the north—as well as a smaller Tharsis volcano in the north, Ceraunius Tholus. Also, small mounds exist in Acidalia and Utopia that may have resulted when lava flowed over them and induced steam explosions.

In 2002, to the delight but not the surprise of scientists, the Mars Odyssey spacecraft discovered subsurface ice poleward of 60 degrees latitude in both the north and the south. The spacecraft detected the ice through its hydrogen, as follows. When spaceborne charged particles called cosmic rays hit Martian ground, they release neutrons. Hydrogen slows neutrons down—that's one reason nuclear reactors use water—and the Odyssey spacecraft saw an excess of slow neutrons, implying the existence of hydrogen and thus of water ice. Cosmic rays also stimulate hydrogen to emit gamma rays, which the spacecraft likewise detected. The ice is no more than a few feet beneath the surface—otherwise, the spacecraft couldn't have detected it—so a thirsty Martian polar bear wouldn't have far to dig.

At times past, Martian ice may have congregated into glaciers. Modest changes in Earth's rotational and orbital parameters are thought to drive its ice ages, so the wilder Martian oscillations described in AIR could have produced far worse. In the early 1990s, Jeffrey Kargel of the U.S. Geological Survey and Robert Strom of the University of Arizona cited several

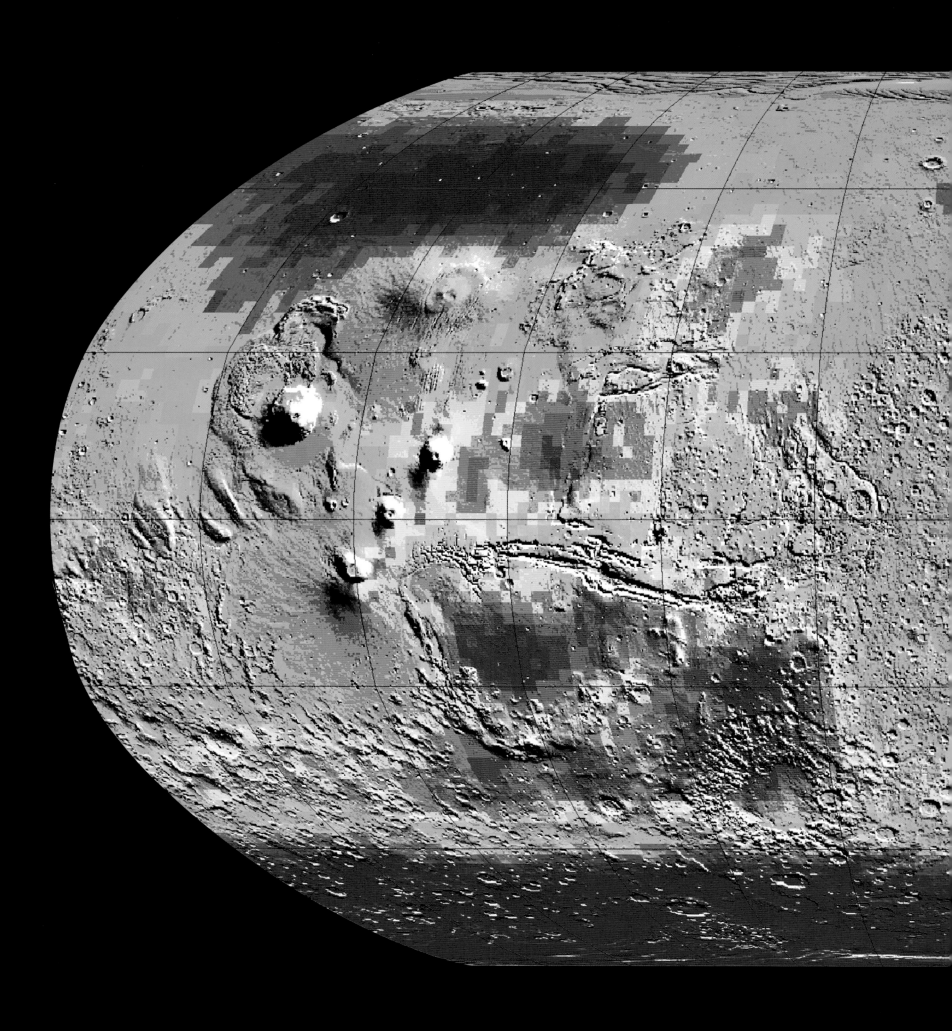

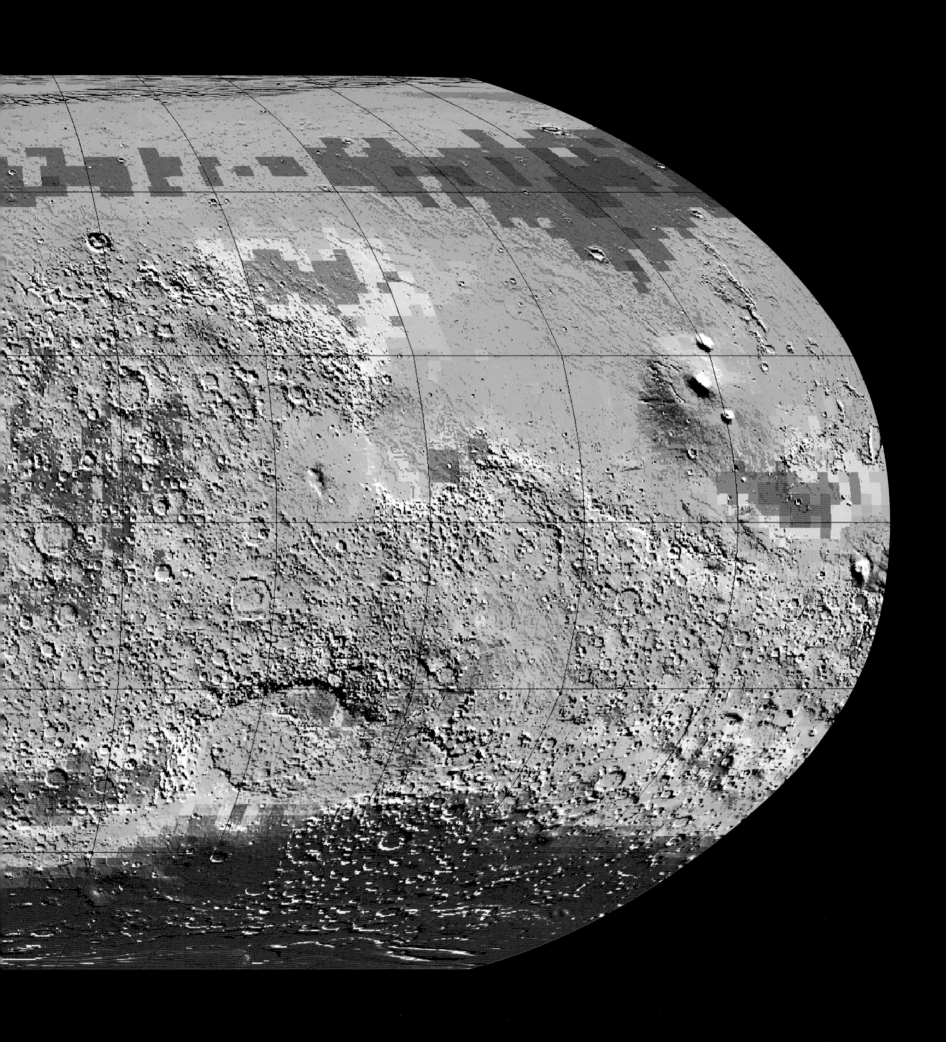

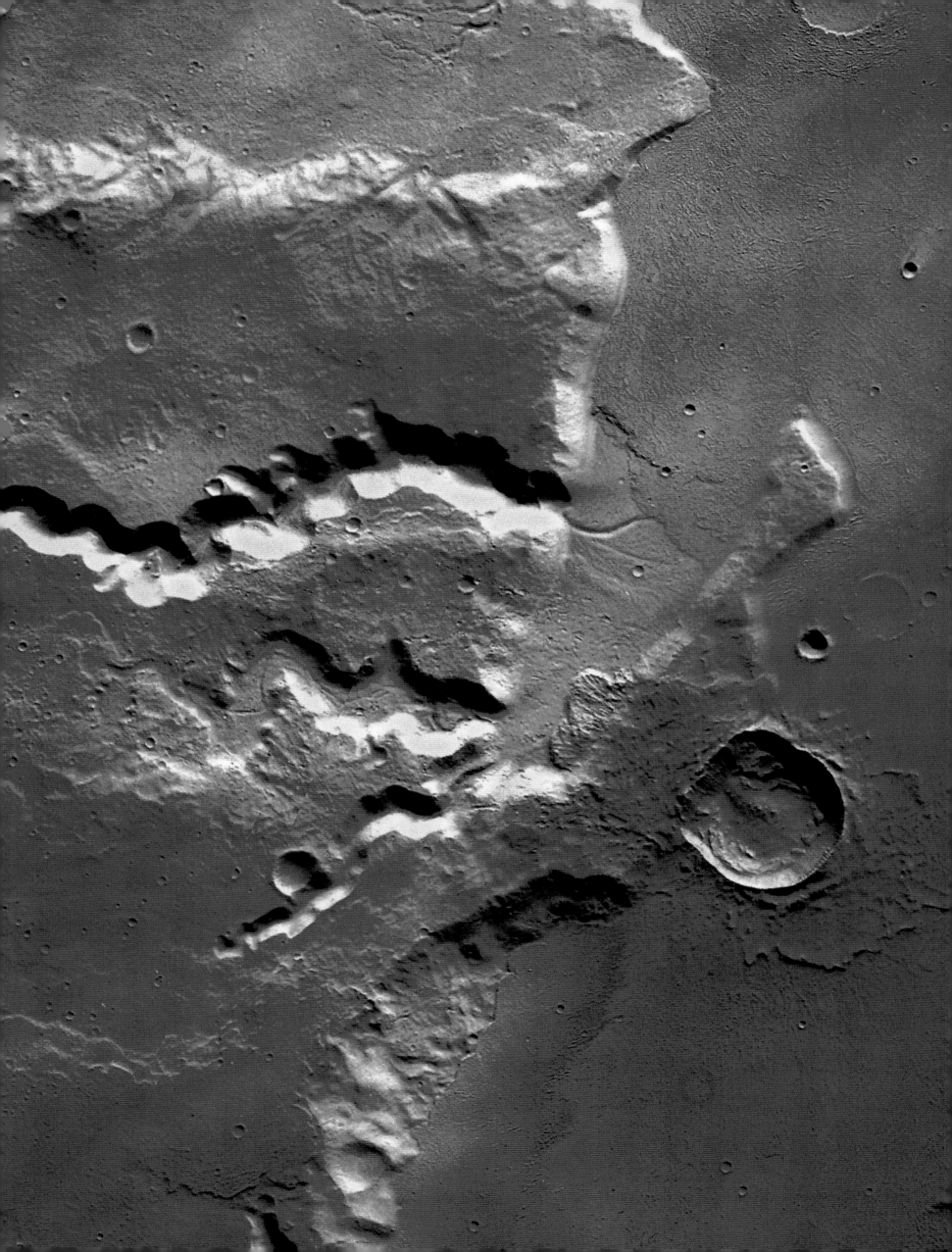

Martian features pointing to past glaciation: for example, eskers and moraines in the northern plains and the southern impact basin Hellas. Eskers are narrow, sinuous deposits left by a glacier stream; moraines are dirt and stones scraped up by a glacier. In addition, the Mars Global Surveyor's laser altimeter confirmed previous indications that Hellas's southwestern rim—close to the south pole—is also the smoothest, perhaps because glaciers from the pole wore it down. Although glaciers connote ice and cold, they demand a warmer, wetter Martian climate—because glaciers on Earth, at least, require snow to grow. Furthermore, Kargel and Strom suggest that some glaciation occurred as late as the Amazonian, implying that warm and wet conditions persisted, at least occasionally, into the most recent period of Martian history.

Water Loss

ALTHOUGH some of the water that flowed on ancient Mars survives as ice at high latitudes, much water escaped the planet altogether. Once water vapor floats high into the air, it is vulnerable to sunlight, which can tear the hydrogen from the oxygen. The hydrogen rises and escapes the small planet's gravity, while the oxygen either escapes or, ever eager for company, combines with the rocks below.

Scientists can try to quantify how much water Mars has lost to space, because the escape alters the proportions of the remaining hydrogen isotopes. The

FACING PAGE: The largest crater in this view of East Mangala Vallis is a splosh crater. It probably formed when an asteroid hit mud.

Much water escaped the planet altogether

lighter and more common hydrogen isotope, hydrogen-1, has one proton; hydrogen-2, twice as heavy, has one proton plus one neutron. Hydrogen-2 is usually called deuterium, after the Greek word for "second," *deuteros*. Deuterium can appear wherever ordinary hydrogen does. For example, most water is H_2O, but some is HDO and even D_2O. About 1 in 7,000 hydrogen atoms in terrestrial water is deuterium, which works out to a deuterium-to-hydrogen ratio of 0.00015, or 1.5×10^{-4}.

The more water that Mars lost to space, the higher its deuterium-to-hydrogen ratio should be—because the light hydrogen escapes more readily than the deuterium. In 1987 the University of Hawaii's Tobias Owen and his colleagues first detected HDO vapor in the Martian atmosphere. Mars has a deuterium-to-hydrogen ratio five times higher than Earth's, indicating that much of the red planet's water has indeed fled into space.

In 2000, however, Laurie Leshin of Arizona State University reported that ancient minerals in the Martian meteorite QUE 94201 have a deuterium-to-hydrogen ratio only twice that of terrestrial water. This deuterium-to-hydrogen ratio may therefore reflect the one that Martian surface water had billions of years ago. By comparing the ancient and modern ratios, scientists think that Mars lost roughly two thirds of its original water.

How did the Earth and Mars originally acquire their water? Up to the late 1990s astronomers thought terrestrial water came from comets, the icy bodies that sport beautiful tails when their ice vaporizes in the Sun's heat. If so, comets should have the same deuterium-to-hydrogen ratio as terrestrial water. Astronomers first measured a comet's deuterium-to-hydrogen ratio in 1986, when Halley's Comet passed Earth. Surprisingly, the comet had twice the terrestrial deuterium-to-hydrogen ratio. Scientists initially dismissed the comet as a fluke, but in 1996 and 1997, when Comets Hyakutake and Hale-Bopp sailed by, they also had twice the terrestrial deuterium-to-hydrogen ratio. Thus, comets did not give Earth most of its water. Instead, it came from the asteroid-like planetesimals that built the Earth.

> # Comets Hyakutake and Hale-Bopp had twice the terrestrial deuterium-to-hydrogen ratio

For Mars, the story is different. As the Martian meteorite reveals, the water of ancient Mars matched the cometary deuterium-to-hydrogen ratio, so the red planet may have received its surface water from comets hitting the planet after it formed. The planetesimals that built Mars presumably had a lower deuterium-to-hydrogen ratio, like Earth's, but because of the red planet's less vigorous geological activity, this interior water hasn't come to the surface the way it has on the Earth.

Early Mars

ANCIENT MARS was wetter and probably warmer than present Mars: rivers flowed, the erosion rate was higher, the deuterium-to-hydrogen ratio indicates more water, and other isotopic ratios, discussed at the end of AIR, indicate more atmosphere. The rivers do not demand rainfall, but if rain ever did fall, the Martian surface must have risen above the freezing point of water. How could a cold desert world like Mars have once been warmer?

Actually, ancient Earth poses a similar puzzle. The present Earth huddles close enough to the Sun to be warm and wet, but the young Sun was 30 percent fainter than it is now. That's because the Sun shines by converting hydrogen into helium at its center; helium is denser than hydrogen; so as the Sun's center fills with helium, it gets denser and must burn brighter, in order to exert a greater outward force to balance gravity's greater inward force. Therefore, when the Sun had just begun to tap its central hydrogen supply, it shone more faintly. As a result, for the first 2.5 billion years of its life, the Earth should have frozen. It didn't. That's because the ancient atmosphere consisted not of nitrogen and oxygen, as it does now, but of carbon dioxide and water vapor, greenhouse gases that warmed the ancient world. Although rainfall eroded rocks and thereby removed atmospheric carbon dioxide, volcanoes shot it back into the air. Eventually, over

FACING PAGE: Deuterium levels in Comets Halley, Hyakutake, and Hale-Bopp (shown here) suggest that comets did not give Earth most of its water.

billions of years, most carbon dioxide did depart the air, and now the equivalent of a 60-bar carbon dioxide atmosphere is locked in carbonate rocks like limestone. When the Earth transitioned from its carbon dioxide atmosphere to its present nitrogen-oxygen one, it probably suffered the worst ice age of its life. However, had the carbon dioxide remained in the air, the Earth today would resemble Venus, whose 93-bar atmosphere is 96.5 percent carbon dioxide and bakes at 860 degrees Fahrenheit.

And Mars? Could its volcanoes have pumped enough carbon dioxide skyward to warm the planet above freezing? Perhaps not. Strangely, as James Kasting of Pennsylvania State University noted in 1991, carbon dioxide has its limits. At the red planet's cold temperature, Kasting found that carbon dioxide in a dense atmosphere would condense high in the atmosphere, thereby warming the air there; much of this warmth would escape into space rather than help the surface below. Furthermore, carbon dioxide clouds would probably cool the planet by scattering sunlight away from Mars. Present models suggest that even if the ancient Martian atmosphere had 1 bar of carbon dioxide, the young Sun's lesser luminosity would have prevented the planet from exceeding a frigid −50 degrees Fahrenheit.

How, then, did Mars get warm enough to spawn rivers? Perhaps other greenhouse gases helped out. For example, volcanoes also emit sulfur dioxide, which is more potent than carbon dioxide— even a small amount significantly boosts the temperature. Sulfur may make up more of Mars than Earth, because sulfur has a fairly low melting point that would have favored its incorporation into a world born at the red planet's greater distance from the Sun. And, as noted in EARTH, the Viking landers did find the Martian soil sulfur-rich. Sulfur dioxide doesn't linger long, however. It vanishes from the air just ten years after the volcanoes shut off.

Other possible greenhouse gases include ammonia and methane. Like sulfur dioxide, both are more powerful than carbon dioxide. For example, according to a model that Kasting reported on in 2003, a mere 0.003 bar of methane would have raised the Martian temperature by about 50 degrees F., though that's still not enough to bring it above freezing. Furthermore, both gases have problems. Ultraviolet sunlight tears ammonia (NH_3) into nitrogen and hydrogen, which escapes into space, and converts methane (CH_4) into more complicated hydrocarbons. Moreover, producing methane may require life, so if methane is the solution, there may be a catch-22: to be warm the planet needs life, but to have life it needs to be warm.

However, more than just the atmosphere warmed ancient Mars. Even if the air lacked exotic greenhouse gases such as ammonia and methane, other factors, plus the carbon-dioxide-and-water-vapor atmosphere, could have boosted the young planet above freezing. For example, the formation of the iron core soon after the planet's birth released so much heat it likely melted the surface, if it wasn't already molten. Volcanoes not only emitted greenhouse gases but also heated the rocks near them. Impacting objects heated the surface and also melted rock, liberating carbon dioxide from carbonate rocks. More quietly, the radioactive decay of underground potassium, thorium, and uranium warmed the world. Taken together, all these processes may have given Mars a brief taste of Earth, when rivers flowed, lakes dotted the surface—and life arose.

MARS FASCINATES not only because it is a near neighbor of Earth but also because it may answer the question of whether the universe teems with life. From its original endowment of elements ancient Earth somehow spawned simple life, which over billions of years evolved into more complex life, some of which ultimately acquired sufficient intelligence to probe the other planets.

Contrary to popular belief, however, no other planet—even one blessed with earthly warmth, sunshine, and seas—need have done the same. After all, planets that don't spawn life don't give rise to creatures who contemplate their nonexistence. Thus, living beings may be biased by their very existence into thinking that good planets like their own inevitably give rise to good life.

In fact, problems crop up every step of the way. Consider the first step, the rise of the first life on Earth. Nonscientists often scoff at microscopic alien life—it's boring, they say—but even today most terrestrial life is microscopic; we macroscopic creatures are the minority. Furthermore, microscopic life is vastly more complicated than the chemicals that presumably led to its creation, which biologists tend to appreciate more than astronomers. As a result, the chance that a good, warm, wet planet shuffles its chemicals in just the right way to spawn life may be abysmally low. No problem, you may think. After all, the observable universe has something like 10^{22} planets. Surely *some* of these have life. But what if the chance of life arising on a planet is only 1 in 10^{80}? Then every planet in the observable universe except Earth is lifeless.

The answer to this question may reside in the rocks of Mars. If Mars independently gave birth to life, no matter how simple, then the chance of life arising on a good planet is hardly 1 in 10^{80} but instead closer

Mars may answer the question of whether the universe teems with life

to 100 percent. Thus, innumerable worlds throughout the cosmos surely did the same. Scientists can then argue about how likely the next two steps are, the evolution of simple life into complex life, and the evolution of complex life into intelligent life. The discovery of indigenous life on Mars would therefore revolutionize a question that philosophers, theologians, and scientists have argued about for centuries.

Even on Earth, however, evidence for ancient life is scant, both because the Earth is so active and because microbes leave more subtle imprints than larger creatures. The oldest terrestrial fossils, in Western Australia, are 3.5 billion years old, so they formed 1.1 billion years after the Earth's birth and 300 million years after the heavy bombardment ceased. These fossils are stromatolites: when microbes on a lake bed get covered with dirt, they migrate upward, toward the light, and disturb the dirt, which eventually solidifies into a layered dome.

Stromatolites may also exist on Mars. "The question is not the truth, but the difficulty of proving it," Australian scientist Malcolm Walter wrote in 1999. "And when it comes right down to it, this is what is going to happen in the search for life on Mars.

We may well find stromatolite-like objects there, but how do we get rid of that unwanted appendage '-like objects'?"

On Earth, in lieu of fossils, scientists have found tentative evidence for even older life. Rocks in Greenland, 3.8 billion years old, contain enhanced carbon-12, the carbon isotope that living beings prefer. If Martian life had the same isotopic taste, then the carbon in the red planet's rocks may provide evidence for ancient life, even without any fossils.

Biologists now realize that terrestrial life can survive and even thrive in hostile conditions, which ups the odds for life on Mars and other planets. For example, so-called thermophile ("heat-loving") bacteria can exist in boiling water. Methanogenic bacteria live in the dark and create methane from hydrogen gas and carbon dioxide. Halophile ("salt-loving") bacteria live in extremely salty places. Anaerobic ("without air") bacteria die if exposed to oxygen. These examples provide hope that some forms of life may exist on a world as hostile as modern Mars.

Because Mars is so cold, the best earthly analogue may be Antarctica, where terrestrial life once again exhibits its adaptability. The McMurdo Dry Valleys are frigid deserts where little precipitation falls and the average annual temperature is around 0 degrees F.; in the summer, the temperature rarely climbs above freezing. Yet life exists. Just inside Sun-facing sandstone rocks lichen and bacteria bask in the warmth of the dark rock, capturing enough stray light to perform photosynthesis. During rare snowfalls, snow gets trapped inside the rocks and on warm days becomes liquid water. Elsewhere in the Antarctic dry valleys, the water-bathed floors of ice-covered lakes nourish algae and other organisms, and the ice lets in sufficient light for photosynthesis. Perhaps Hellas and Argyre were once similar.

Terrestrial life therefore proves capable of adapting to extremes. One thing it can't adapt to, though, is a lack of liquid water. Even on Earth, in the dryest parts of the Antarctic dry valleys, 10 to 15 percent of the soil samples are sterile. If all Martian water today is frozen, the red planet is likely a dead planet. On the other hand, if life ever arose on Mars, and if at least some of the planet's water has remained liquid, then the tenacity of life suggests that Martian beings may survive to this day.

Of course, these conclusions assume that living beings on Mars resembled their terrestrial peers. Ironically, one reason scientists want to study life on Mars is to see whether it differed from life on Earth—but in the search for such life, scientists have only the example of the Earth's life to follow.

Scientists often assume that life arose on Earth only after the heavy bombardment of asteroids and comets ended, 3.8 billion years ago. After all, the impacts devastated the Earth's surface, blasting rock and perhaps tearing away part of the atmosphere. But such thinking may be too conservative. Just as thermophiles survive heat that would kill most creatures and anaerobic bacteria perish in the oxygen that sustains human life, perhaps exotic organisms—call them "impactophiles"—thrived on Earth and Mars during the 800 million years of the heavy bombardment. These impactophiles would have relished the chaos, savored the extremes of heat, perhaps even feasted on the melted rock—and died out when their planet settled down to a more sedate existence.

To discover signs of past Martian life, where on the red planet should astronauts look? Since no one knows how life arose on Earth, the wisest strategy would cast a broad net, examining diverse geological settings rather than focusing on just one or two. If Mars once had a northern ocean, life might have started on its floor. If Hellas, Argyre, and other craters once hosted lakes, their deposits may preserve ancient fossils. The layered walls of the Valles Marineris canyons may do the same, as may the hematite-rich regions of Meridiani Planum. The riverbeds, where water flowed for prolonged periods, are also ideal. The polar caps still have water today, albeit frozen, and might have once nourished Martian life, especially during climate cycles that brought sunshine and warmth to the poles. Volcanic heat may have triggered life, so Tharsis and Elysium are logical hunting grounds. In recent years biologists have found some evidence that the first life on Earth preferred high temperatures, so Martian hot springs and geysers akin to Yellowstone's Old Faithful should be searched. No hot springs or geysers, past or present, are known on Mars, but they are difficult to spot from space, and they may exist in the volcanic provinces. The recently discovered gullies, if carved by liquid water, even offer the hope of finding life that survives to this day.

If Martian fossils exist, they may predate any known life on Earth, because Mars has erased less of its ancient terrain. The oldest evidence for life on Earth—the carbon isotope anomaly in Greenland—dates back to the end of the heavy bombardment, but the Martian record of life could conceivably extend well into the heavy bombardment. It's even possible that life developed on Mars *before* it did so on Earth—perhaps the young Earth was too hot for life, whereas more distant Mars was just the right temperature. Should no evidence of Martian life be found, the planet may nevertheless preserve the chemical steps that led to life on Earth.

One complication exists, however. The Martian meteorites demonstrate that material has moved from Mars to Earth. It can also move from Earth to Mars. Because of the Earth's greater gravity, this is harder, but it's surely happened. If terrestrial microbes sur-

vived the trip, they could have contaminated Mars. Unfortunately, then, if astronauts do discover past life on the red planet, scientists may never know whether it arose there or on Earth, thwarting their effort to address the issue that motivates much Martian research—determining how likely life is to arise on a mild planet. Then the focus may shift beyond the asteroid belt to Jupiter's icy moon Europa, whose distance better protects it from terrestrial contamination. Scientists can hope that any Martian life possessed unique chemical features that distinguish it from terrestrial life, but even then, how would they know that this unique chemistry had not also existed in ancient terrestrial beings that left no fossils?

Possibly the opposite happened: life first arose on Mars and then migrated to Earth on the backs of meteorites. If so, to see Martian life, all we need do is look in the mirror.

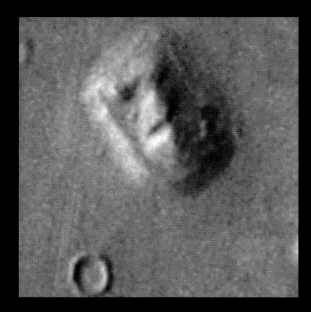

ABOVE AND FACING PAGE: Mars loses face: above is the Viking image of the face on Mars; on the facing page, the high-resolution image from Mars Global Surveyor.

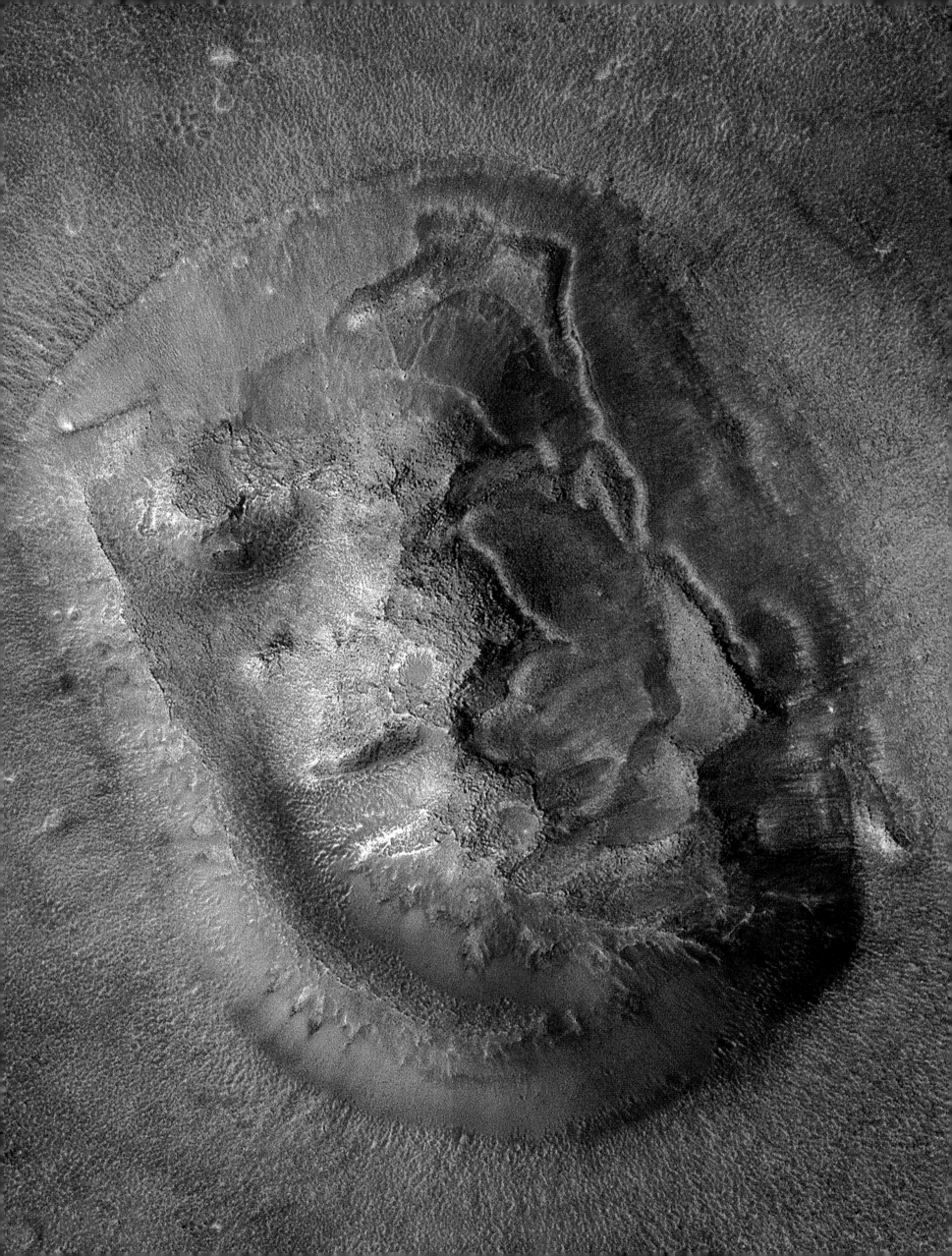

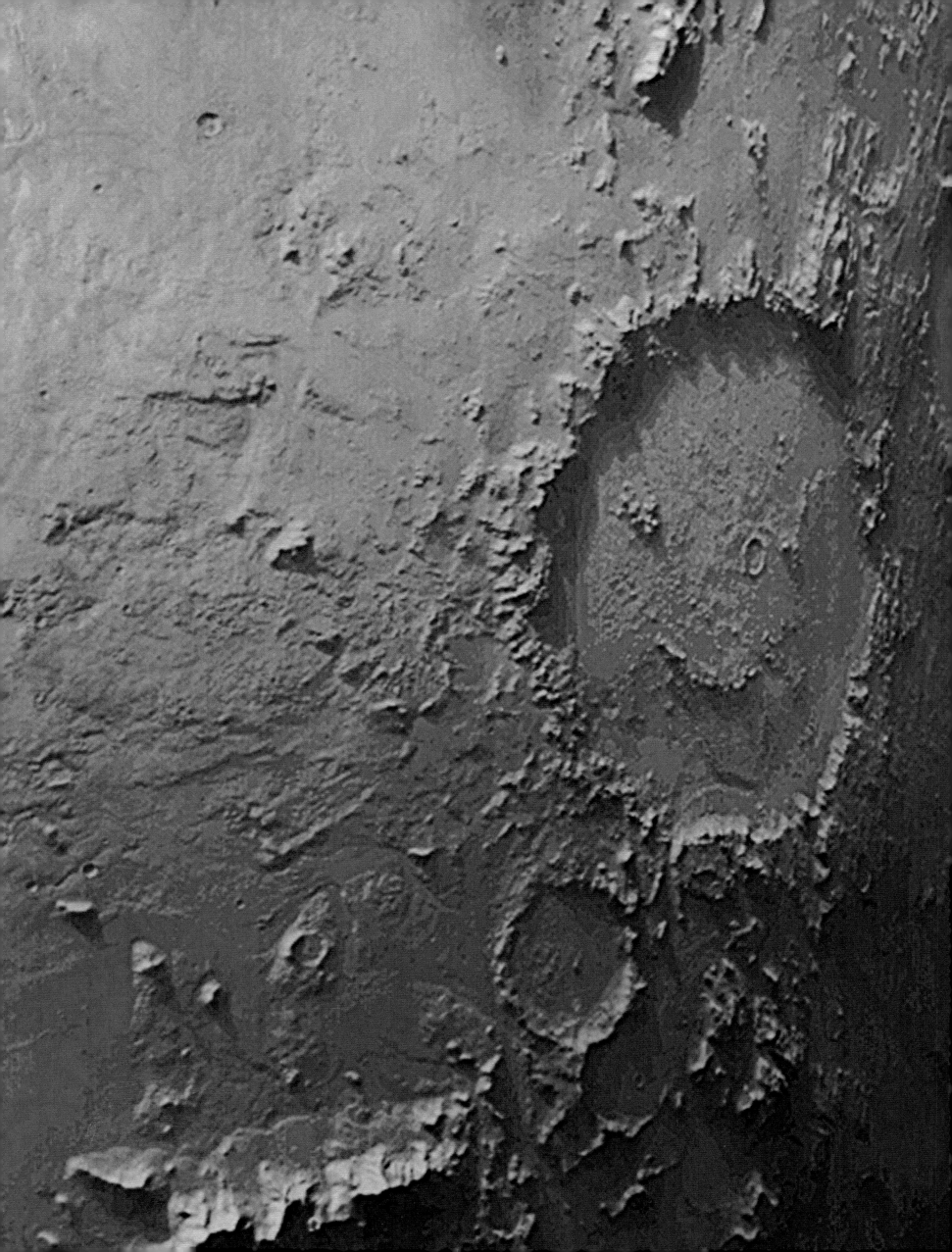

Fossils in Martian Meteorites?

IN 1996 scientists led by David McKay of NASA's Johnson Space Center in Houston claimed to have detected fossils of microscopic life in the 4.5-billion-year-old Martian meteorite ALH 84001. Even then, most scientists were skeptical, and they have become only more so since.

McKay's team identified tiny wormlike structures that resembled bacteria; organic compounds called polycyclic aromatic hydrocarbons, or PAHs; iron sulfide and iron oxide near each other, suggestive of organic activity, inside globules of carbonates; and, most convincingly, crystals of magnetite that only bacteria were known to produce.

Despite this evidence, trouble erupted. First, the putative bacteria were much smaller than what most biologists consider bacteria. Second, the PAHs don't demand living organisms—PAHs also exist in meteorites from presumably lifeless asteroids. Third, iron sulfide and iron oxide can appear close together even in the absence of life. Fourth, the magnetite crystals may indeed have formed when Martian bacteria oriented themselves in the ancient magnetic field, but they may have a nonbiological origin that scientists have not yet discovered.

FACING PAGE: Galle Crater, in the Argyre impact basin, smiles for the camera. A fortune surely awaits the first person to claim it was built by intelligent beings. North is up.

Circumstantial evidence weighs against the Martian fossils

Circumstantial evidence also weighs against the Martian fossils. Because of the difficulty of finding microfossils even on the Earth, noted the University of Arizona's Ralph Lorenz, "It would be an incredible stroke of luck if one of the handful of rocks we have from Mars just happens to be teeming with obvious fossils."

Face-to-Face

WHILE the possible Martian fossils sparked a legitimate scientific debate, the purported face on Mars did not. In 1976 the Viking 1 orbiter spotted a feature on the Martian surface that resembled a human face, the way a passing cloud might, and scientists issued a press release calling attention to the amusing find. Little did they know that a science writer named Richard Hoagland—to his critics, "Hoaxland"—would claim the face on Mars had been built by intelligent beings.

As author Laurence Bergreen quoted one scientist, "This is such a load of crap." When the Mars Global Surveyor scrutinized the face, it found, to no scientist's surprise, a mere jumble of hills and dales.

The History of Mars: When Bad Things Happen to Good Planets

THROUGH the centuries astronomers have transformed Mars from a ruddy orb connoting war into a neighbor world that promises to answer key questions about extraterrestrial life. Early observations showing a terrestrial rotation rate, polar ice caps, white clouds, blue seas, green vegetation, and the infamous canals raised hopes that spacecraft later dashed, revealing a cratered world now cold and dry, its best years long gone.

The red planet may have gotten off to a good start

Nevertheless, careful examination of the four elements of Mars—EARTH, AIR, FIRE, and WATER—demonstrates that the red planet may have gotten off to a good start. The composition of the Martian meteorites indicates that Mars was born hot, like the Earth, from the heat of the formation of its iron core and from the planetesimals pounding the planet. Its early surface was likely molten, and even after the surface solidified, vigorous volcanoes, especially in the Tharsis region, spewed carbon dioxide, water vapor, and other gases into the atmosphere. Plate tectonics—continental drift—may have churned the land and released additional gases. Small planetesimals smashing into Mars vaporized materials, too, contributing still more to the atmosphere.

Greenhouse gases such as carbon dioxide and water vapor trapped warmth from the young Sun and from the surface below. A strong magnetic field, originating in the planet's iron core, guarded the atmosphere from the devastating solar wind. Rivers flowed down Martian hills, lakes likely gathered in craters, and a possible ocean filled much of the northern plains. Somewhere—in a warm pond, beside a hot geyser, near a volcanic vent—chemical reactions may have sparked the first Martian life. These organisms may have further helped warm the planet by releasing methane, an especially potent greenhouse gas.

Unfortunately, Mars was born crippled. First, the planet was 50 percent farther from the Sun than was the Earth, and the young Sun was only 70 percent as bright as its modern incarnation. As a result, ancient Mars received only 30 percent of the sunlight that strikes present-day Earth. Second, Mars was born too small. By rights it should have grown about as large as the Earth; but because of Jupiter's interference, Mars achieved only 11 percent of the Earth's weight.

Consequently, Mars was born with two strikes against it, forcing it to struggle to maintain a mild climate capable of sporting rivers and sprouting life. Because of its small size, Mars was much more vulnerable to the devastating effects of impacting asteroids than were ancient Venus and Earth. Although small impactors delivered gas to the Martian atmosphere, giant impacts blasted air away—though no one knows how much. Also because of its small size, the Martian interior cooled fast. Within a few hundred million years the planet lost its magnetic field. As a result, the atmosphere stood defenseless against the solar wind, which began to cast atoms and molecules in the red planet's upper atmosphere into space. Also because of the cooling interior, the volcanoes began to erupt more sporadically, lessening their supply of gas to the atmosphere just when the planet needed it most to retain warmth. Any continental drift ceased,

The Earth, one hopes, will be more fortunate

to the polar caps, some dove underground. And when the last drop of liquid water vanished from Mars, so did any life. The end result of this tragic tale of planetary

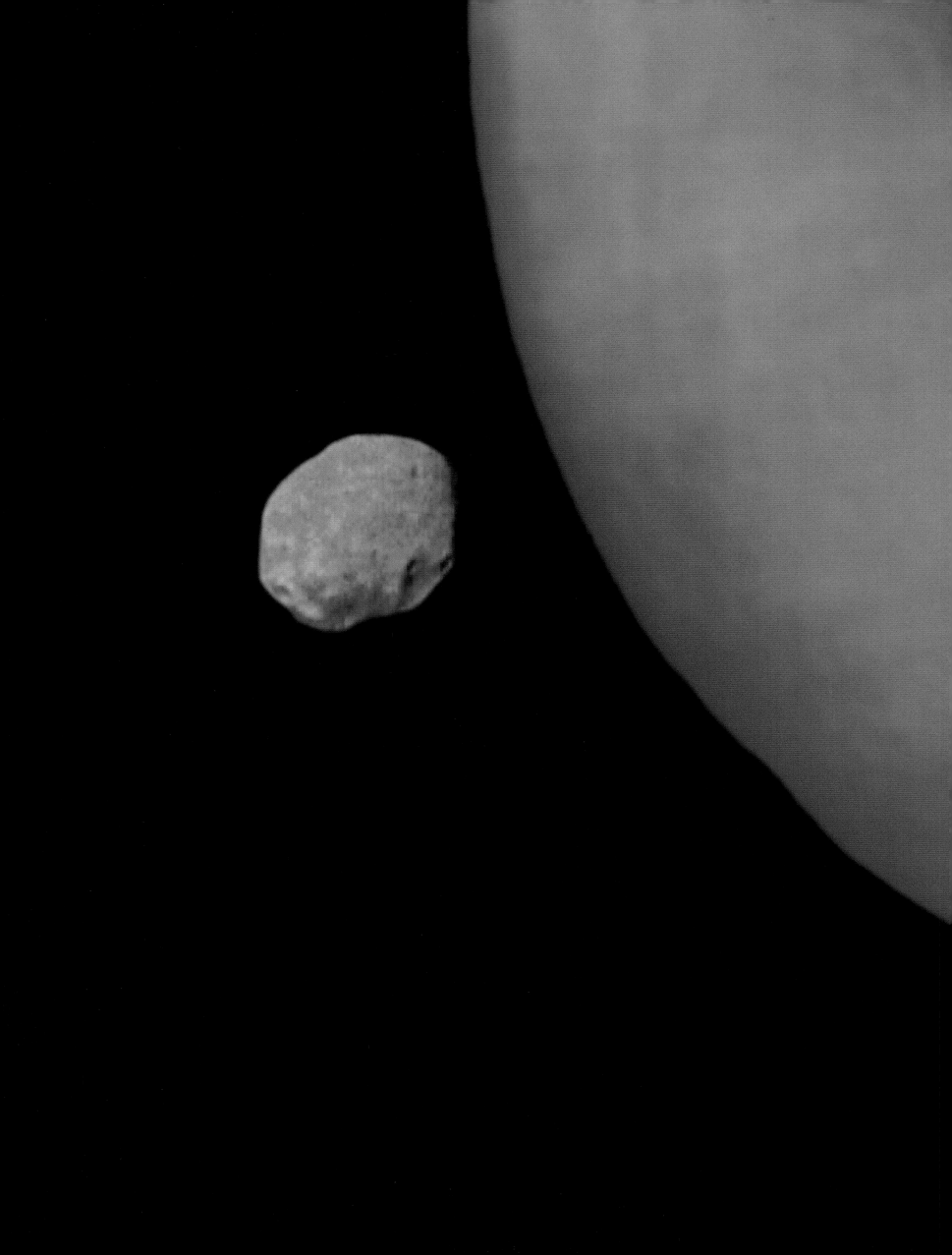

The Moons
of Mars

MARS HAS ATTENDANTS. Two dark satellites scurry about the red planet, Phobos ("fear")

and Deimos ("terror"), fitting names for the war god's companions. Yet the worlds themselves hardly

inspire trepidation: the larger, Phobos, is only 14 miles across, 1/155 the diameter of Earth's Moon, so

you could walk around it in just a day. If you weighed a hundred pounds on Earth, you'd weigh only an

ounce on Phobos. Deimos is even tinier, a mere 8 miles across. The moons resemble asteroids—

they're potato-shaped and crater-pocked—suggesting that they may be former residents of the aster-

oid belt that strayed too close to Mars. Indeed, they are little larger than the asteroid that killed the

dinosaurs. And in some 40 million years, Phobos will come crashing down onto the Martian desert.

FACING PAGE: The dark moon Phobos circles Mars.

Discovery

"A phenomenon solved by modern philosophy and
astronomy." —Jonathan Swift, *Gulliver's Travels,*
Part 3, Chapter 3

THE FIRST fairly accurate description of Phobos
and Deimos appeared in 1726—one and a half centuries *before* anyone ever saw them. Ironically, this
depiction appeared in Jonathan Swift's satire *Gulliver's
Travels,* which followed its hero to fantastic places and
ridiculed scientists along the way. Swift assailed the
presumption that science can lead one to a godlike
stature. He believed this craving for perfection ultimately became a desire to relinquish humanity.
Indeed, Gulliver's fourth and final voyage took him to
a land where horses were wise and people barbarians,
causing him to make "a firm resolution never to return
to human kind."

On Gulliver's third voyage, he visited the flying
island of Laputa and found a band of astronomers con-
trolling the island's flight through the air. These
astronomers far surpassed their European counterparts. They had catalogued three times as many stars
and found ninety-three comets. "They have likewise
discovered two lesser stars, or 'satellites,' which
revolve about Mars, whereof the innermost is distant
from the center of the primary planet exactly three of
his diameters, and the outermost five; the former
revolves in the space of ten hours, and the latter in
twenty-one and an half; so that the squares of their
periodical times are very near in the same proportion
with the cubes of their distance from the center of
Mars, which evidently shows them to be governed by
the same law of gravitation, that influences the other
heavenly bodies." Here Swift tucked another absurd-
ity, for his moons revolved around Mars faster than

the planet spun. Thus, a Martian colonist would see
them rise in the west and set in the east—opposite
from every moon then known.

Astronomers first glimpsed the actual Martian
moons in 1877. The story of their discovery began the
year before, however, with the planet Saturn.
American astronomer Asaph Hall, working at the U.S.
Naval Observatory in Washington, D.C., had discovered a white spot on the ringed planet's globe. This
white spot moved as Saturn spun. By observing the
spot's position, night after night, Hall determined that
Saturn's rotation period was nearly a quarter hour
shorter than the textbooks had said.

Hall wrote, "This discordance . . . set before me
in a clearer light than ever before the careless manner
in which books are made . . . and made me ready to
doubt the assertion one reads so often in the books,
'Mars has no moon.' " As a result, when Mars neared
Earth in 1877—the same year that saw the "discovery"
of the Martian canals—Hall decided to search for
satellites, taking aim with the Naval Observatory's
large new 26-inch telescope.

He knew the task would be difficult. Other
astronomers had looked but found nothing.
Furthermore, Mars is small, so Hall realized that any

FACING PAGE: Phobos bears grooves that radiate from
Stickney, the large crater at upper left.

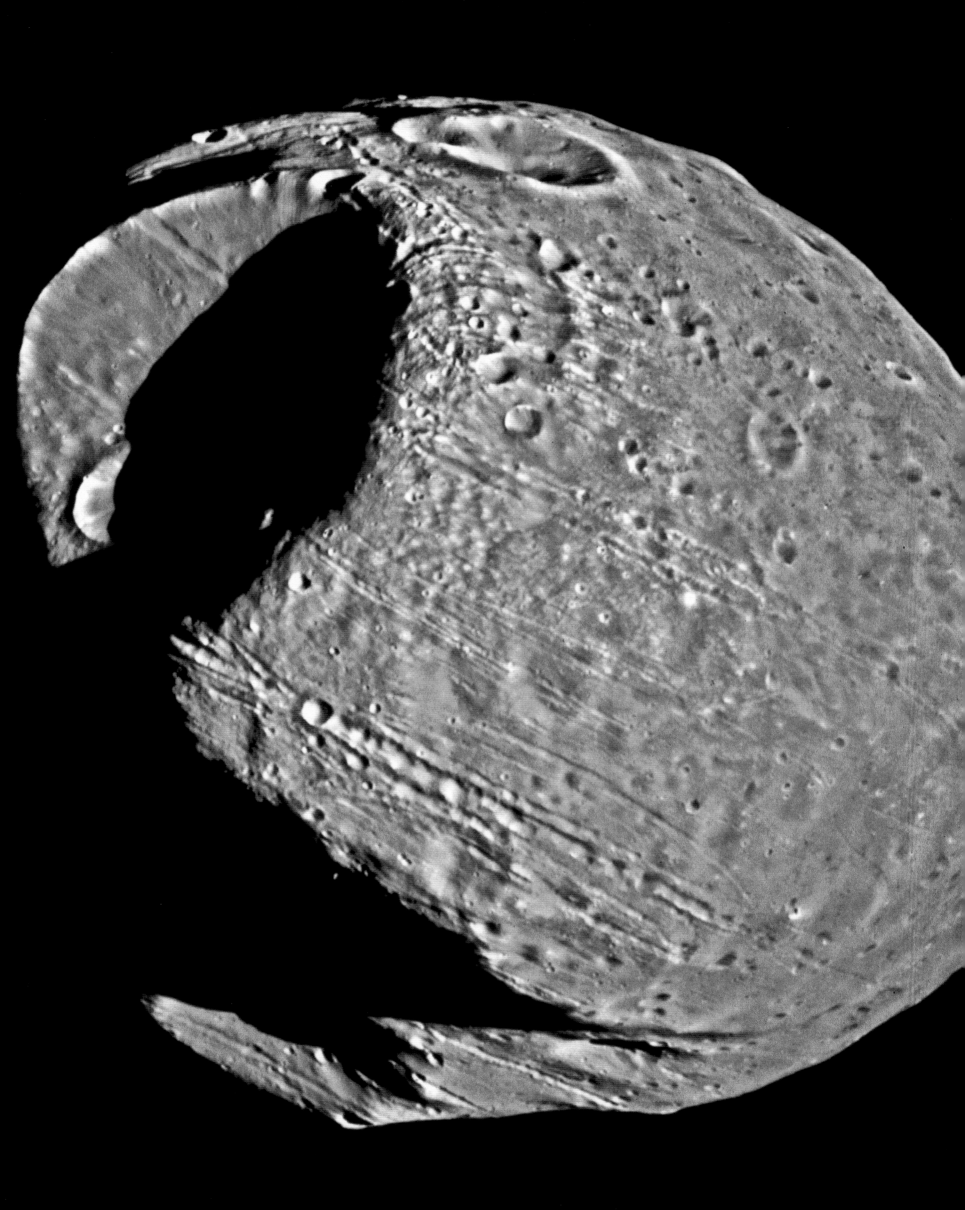

moons must lie close to the planet or else the Sun's gravity would have yanked them away. Trouble was, as viewed from Earth, such close-in satellites would be overwhelmed by the Martian glare. Indeed, as we know today, Phobos and Deimos are only 1/200,000 and 1/600,000 as bright as Mars. In addition, Mars was in the southern sky, as it always is during favorable oppositions, so Australian astronomers were better poised to make the discovery. At one point, after discovering faint points of light near Mars that proved to be only stars, Hall gave up the search, but his wife, Angeline, urged him to persevere.

> Phobos and Deimos
> are only 1/200,000
> and 1/600,000 as
> bright as Mars

To maximize his chances, Hall searched close to Mars but kept the planet—and its glare—just outside the telescope's field of view. As he later wrote, "In the case of the Mars satellites there was a practical difficulty . . . It was to get rid of my assistant. It was natural that I should wish to be alone; and by the greatest good luck Dr. Henry Draper invited him to Dobb's Ferry at the very nick of time. He could not have gone much farther than Baltimore when I had the first satellite nearly in hand."

Hall glimpsed that moon, Deimos, around 2:30 on the night of August 11/12, 1877: "I had barely time to secure an observation of its position when fog from the [Potomac] River stopped the work." Although Deimos is smaller and dimmer than Phobos, it lies farther from the Martian glare and so is easier to see. Still, at that time, Hall did not know if the speck of light he saw was really a moon—it might have been just another background star. Thus, Hall sought to reobserve the object and see whether it moved with Mars, as a moon should; but the weather stayed bad till dawn. The next three nights were cloudy, too. On the night of August 15, an early evening thunderstorm so distorted the air that when the sky cleared, Mars was bobbing around like a pebble beneath turbulent water. Finally, the next night, Hall saw Deimos again. It was a true moon, for it was moving with the planet.

Well after midnight the next night, while Hall was waiting for Deimos to emerge from behind the planet, he discovered the inner moon, Phobos. At first, however, he thought Mars had two or three inner moons, for he saw the moon on opposite sides of the planet during the same night. In fact, just as Jonathan Swift had said, Phobos revolves around Mars faster than the planet spins, so it can appear on opposite sides of the planet in the same night. It revolves three times in a single Martian day—once every 7 hours and

FACING PAGE: Stickney, at upper left, is the largest crater on Phobos, measuring six miles across.

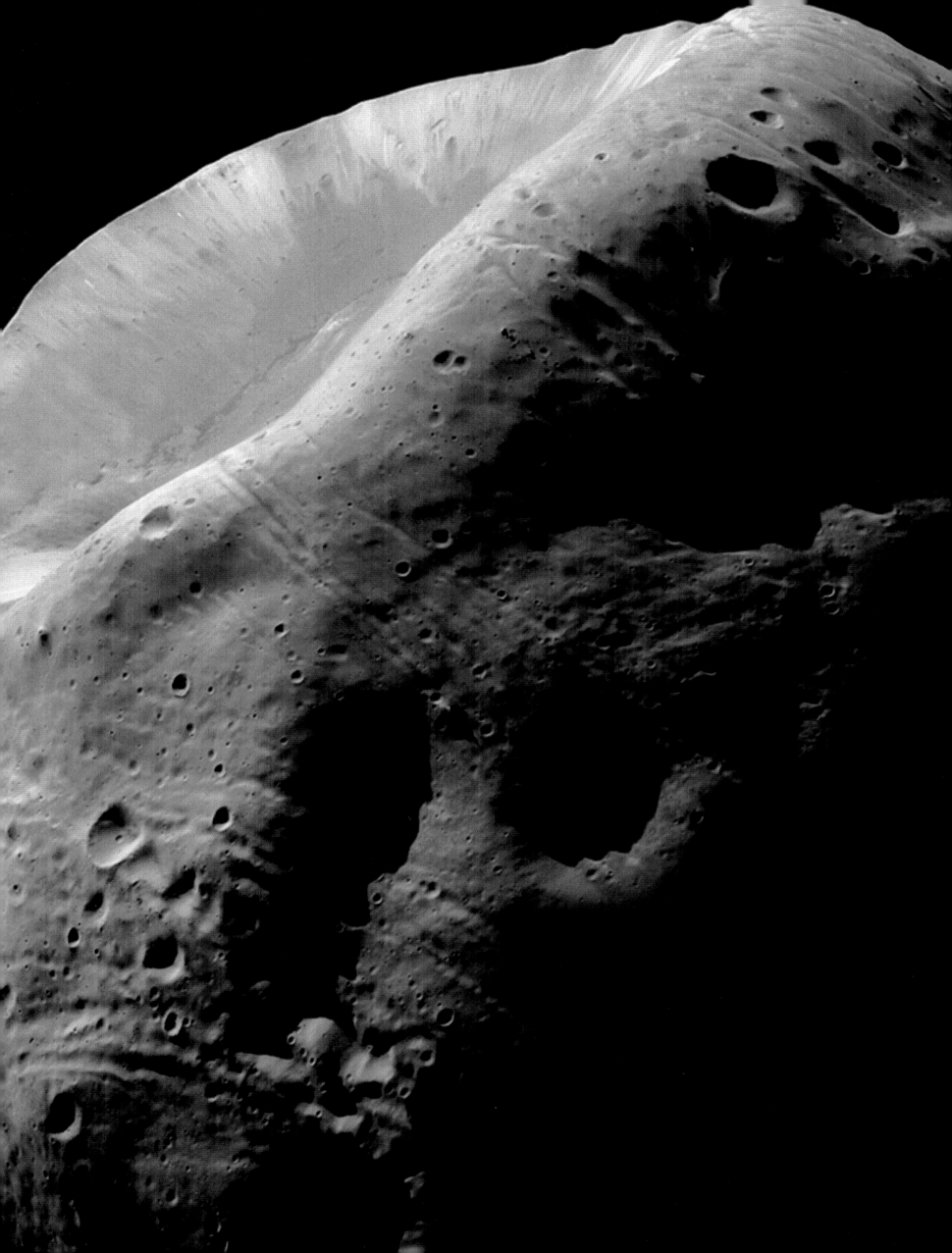

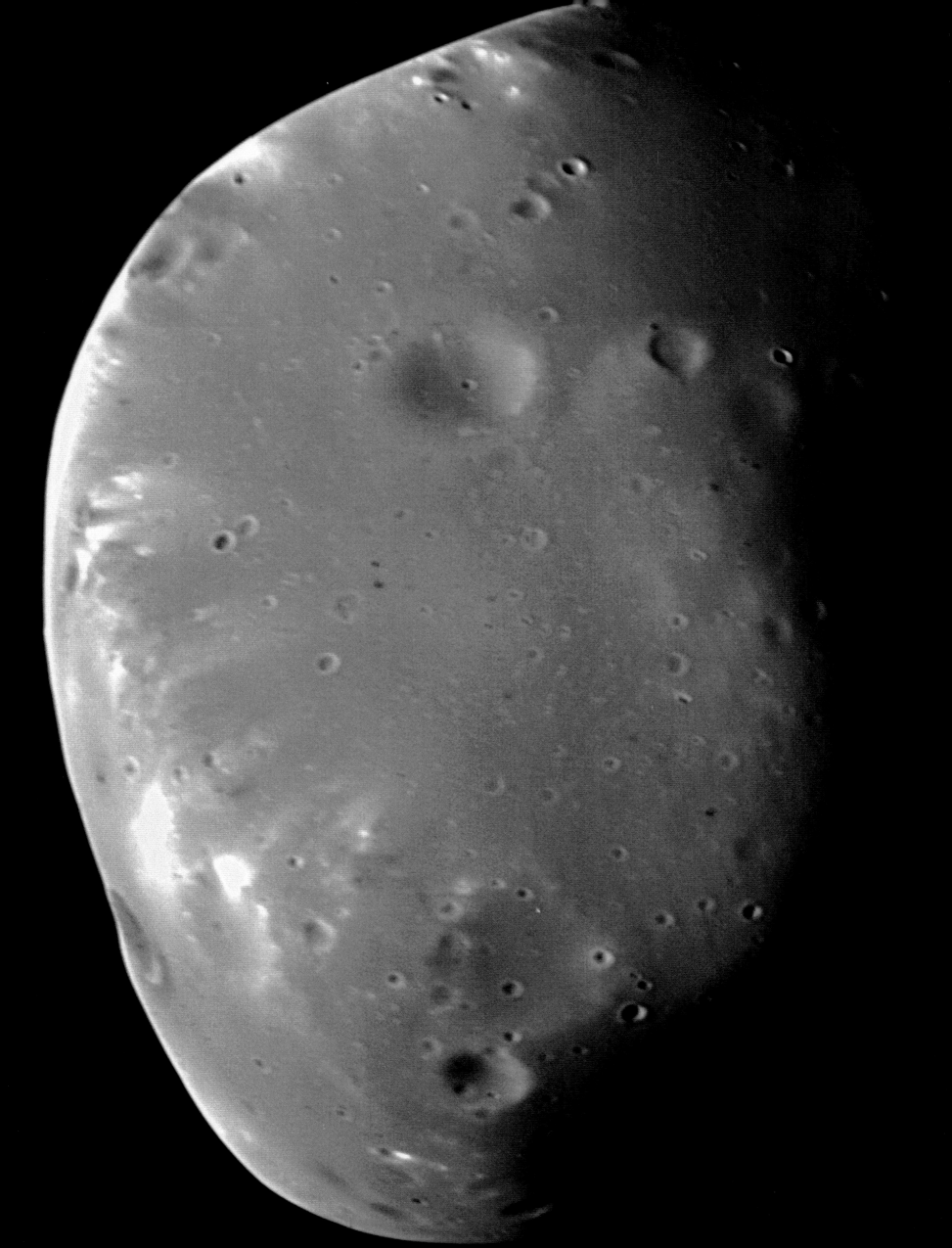

39 minutes—so a Martian colonist would see the faint moon rise in the west, go through its phases in just eleven hours, then set in the east. Dimmer, distant Deimos revolves more slowly, once every 30 hours and 18 minutes, almost six hours longer than a Martian day. Since the moons move under the influence of the red planet's gravity, their motions yielded the first trustworthy estimate of the Martian mass.

Hall named the moons Phobos and Deimos, after the war god's mythological companions. The names were first suggested by Henry Madan, a scholar in Eton, England. Madan's family went on to christen another small world. In 1930 his eleven-year-old grandniece, Venetia Burney, suggested that a newly discovered planet be called Pluto.

So, just as Jonathan Swift had said, Mars has two moons. One of them even revolves faster than the planet rotates. How did Swift know? Actually, although Swift was the first to describe the moons in detail, he was not the first to suggest their existence; Johannes Kepler was. Although Kepler was the great astronomer who worked out the laws of planetary motion and recognized that the planets follow elliptical orbits, he was also a mystic obsessed with astrology and numerology. Kepler thought he saw a mathematical pattern to the planets' moons: Mercury and Venus had none, Earth had one, Mars had an unknown number, and Jupiter had four known moons. Since Mars lies between Earth and Jupiter, Kepler thought its moon count should be intermediate, too.

Kepler assigned two moons to Mars and six or eight to Saturn. By luck, then, Kepler—and Swift—got it right, but Swift probably included the prediction in *Gulliver's Travels* to mock Kepler.

All heroes and heroines lived happily ever after—as landmarks on the Martian moons. In the 1970s scientists named the largest crater on Phobos Stickney, the maiden name of Hall's wife, who had urged him to persevere; the second largest crater, near the south pole, Hall; a ridge in the southern hemisphere, Kepler; and a small crater on Deimos, Swift.

> Since the moons move under the influence of the red planet's gravity, their motions yielded the first trustworthy estimate of the Martian mass

FACING PAGE: Deimos, the smaller Martian moon, looks smoother than Phobos but shares the same odd shape.

Phobos and Deimos

STAND on the side of Phobos facing Mars. Because the moon hovers so close to the Martian surface—closer than Honolulu is to Chicago—the red planet looms over a hundred times larger than the Moon looks from Earth, bathing you in orange light thousands of times stronger than terrestrial moonlight. However, the surface of Phobos is nearly as dark as fresh asphalt. A few miles west of you is the crater Stickney, six miles across, nearly half the satellite's mean diameter. To the south is the crater Hall, four miles across. If you don't like Phobos, you can always leave—by jumping off. That's because the gravitational pull of this small world is so weak that here, on the point facing Mars, the escape velocity is just 7 miles per hour. Drop a pebble from the height of your waist and it would take nearly half a minute to reach the ground.

Because of the weak gravity, neither Phobos nor Deimos is forced to be round. Large worlds, such as the Earth and Mars, are round because their gravity squeezes them into shape. Phobos and Deimos resemble potatoes. For Phobos, the longest dimension, from tip to tip, is 16 miles; the second longest axis, 14 miles; and the smallest axis, 11.5 miles. Deimos has a similar shape but is smaller: 10 by 7.5 by 6 miles. The same side of each moon perpetually faces Mars, just as the same side of the Moon faces Earth. For each moon, the longest axis points toward Mars; the second longest axis points along the moon's orbit; and the shortest axis is perpendicular to both and harbors the moons' north and south poles.

Phobos hasn't even a millionth of the Moon's mass, one reason for its low escape velocity. However, Mars also helps. On the tip facing Mars, not only are you farthest from Phobos' center, but the Martian gravity also tries to pull you off the moon. On the opposite side of Phobos, directly away from Mars, the escape velocity is similar, for the same two reasons—but this time Mars is trying to pull Phobos out from under you. The maximum escape velocity on Phobos is 40 miles per hour, still a far cry from the Earth's escape velocity of 25,000 miles per hour. Although Deimos has the same odd shape as Phobos, its escape velocity varies less—from 10 to 16 miles per hour—because it resides farther from Mars, whose gravity therefore affects it less.

Both moons have suffered countless impacts from space debris. Deimos looks smoother than Phobos, since soil has partially filled the smaller moon's craters. On Phobos, the impact that created the crater Stickney nearly shattered the moon altogether. Grooves up to twelve miles long radiate from the crater, probably caused by the stress of the Stickney impact. Mars may once have had more than two moons but lost them to impacts even more catastrophic than the one which excavated Stickney.

FACING PAGE: This topographic map of Phobos colors high altitudes red, middle altitudes yellow, and low altitudes blue. Here, the leading hemisphere of Phobos plows into space as the small moon circles Mars. Mars is to the right, its surface 190 feet away on the same scale. The large green and turquoise crater near the right tip is Stickney, and the intense red immediately left of Stickney marks part of the crater's rim. North is up.

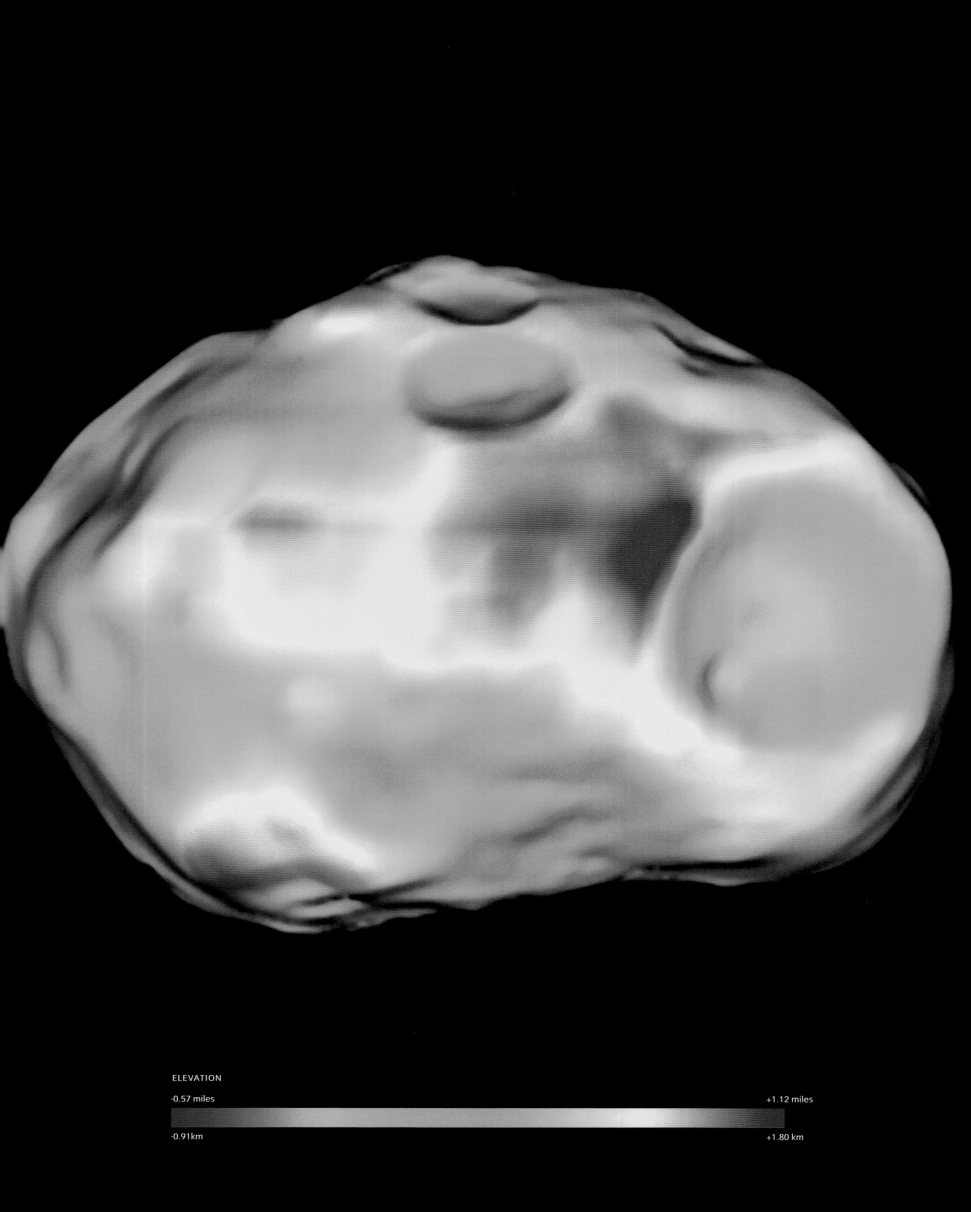

ELEVATION

-0.57 miles   +1.12 miles

-0.91km   +1.80 km

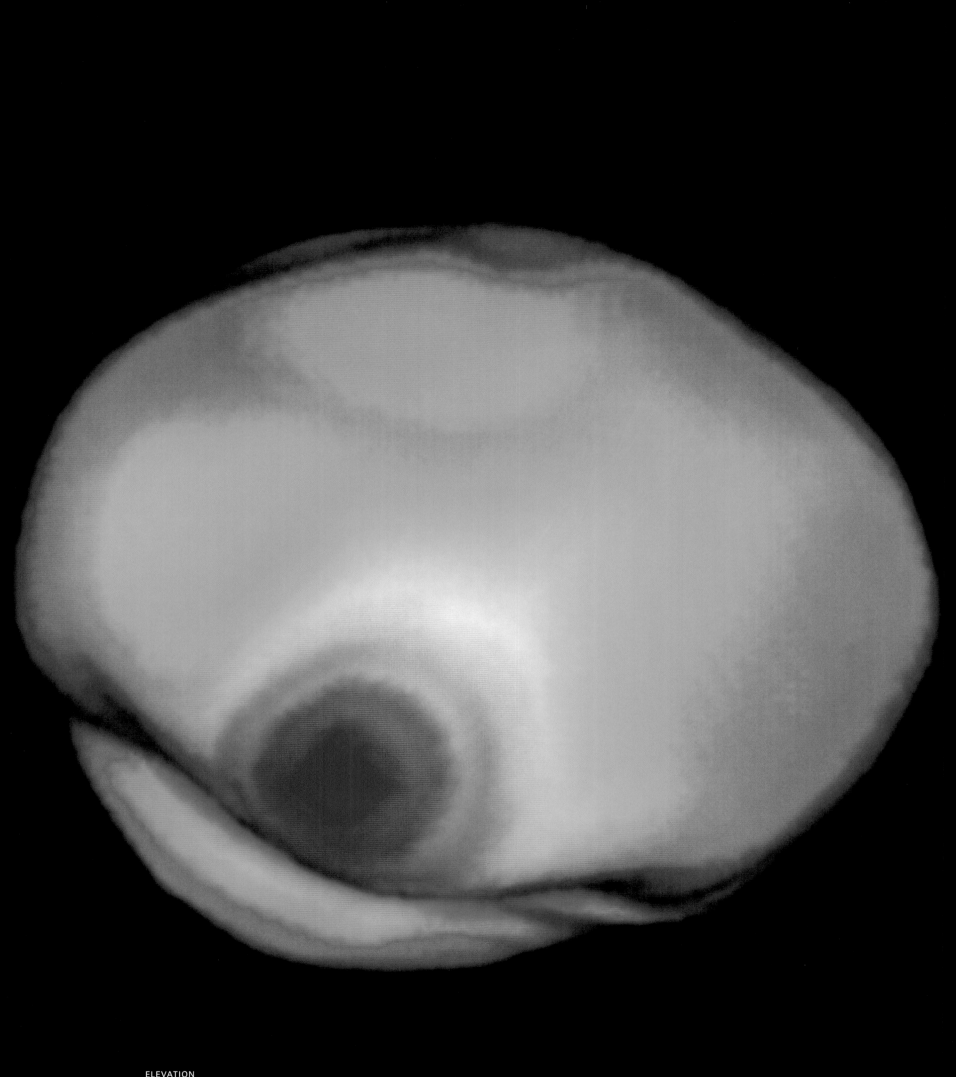

ELEVATION

-0.12 miles   +1.11 miles

-0.20 km   +1.79 km

Because of the moons, and their low escape velocities, Mars should be encircled by two rings of dust. The dust rings arise because impacts on Phobos and Deimos eject material that the satellites' weak gravity can't retain. These rings are so faint, though, that they have yet to be seen.

Phobos and Deimos are dark, reflecting only 7 percent of the light that strikes them. Both are lightweight, less than twice as dense as water, half as dense as Mars. Their darkness and low density suggest that they are made of carbonaceous chondritic material, a dark substance that pervades the outer asteroid belt. Indeed, with their dark color and low density, Phobos and Deimos resemble dark asteroids that astronomers have classified type C, for carbonaceous. In 1991 and 1993 the resemblance between the moons and asteroids became even more apparent when the Jupiter-bound Galileo spacecraft sailed through the asteroid belt and returned the first close-up asteroid images. The two asteroids that Galileo flew past, Gaspra and Ida, proved to be potato-shaped and crater-pocked, like Phobos and Deimos. Ditto for the asteroid Mathilde, which the NEAR Shoemaker spacecraft flew past in 1997, and also Eros, which the spacecraft landed on in 2001.

The Martian moons may actually be former asteroids that the red planet captured. After all, Mars resides beside the asteroid belt, and its giant neighbor on the asteroid belt's other side has undoubtedly done its share of asteroid snatching. Jupiter's outer moons travel around the giant planet on highly elliptical and inclined orbits, and several revolve backward, all signs that they did not form with the planet. Saturn, Uranus, and Neptune also have captured moons.

Trouble is, although Phobos and Deimos look like captured moons, they don't follow the paths that such worlds should. Both moons are close to Mars, not far away; their orbits are nearly circular, not highly elliptical; they revolve in nearly the same plane as the Martian equator, not at great angles to it; and they revolve in the same direction that Mars spins, not backward. As a result, some scientists think that the moons formed with Mars, and from the same material. But then it's hard to understand why they look so different from their master.

Oddly, astronomers are observing Phobos during the last 1 percent of its life. In 1945, Bevan Sharpless, working at the same observatory as had the moon's discoverer, found that the orbit of Phobos is decaying. He did so, strangely enough, by discovering that Phobos was speeding up. In the sometimes paradoxical world of orbital dynamics, when a moon falls inward, it has to move faster, to keep from crashing into the planet. In like fashion, a student whose grades are slipping ought to work harder to avoid losing further ground. Phobos is 20 degrees farther along its orbit than it would be if its orbit had not shrunk since its discovery.

In 1959, Soviet astronomer Iosef Shklovskii seized upon this observation to suggest that Phobos and its mate, Deimos, were not natural moons but

FACING PAGE: This topographic map shows the side of Deimos that faces away from Mars. The red bull's-eye is part of the rim of a large crater that scars the moon's south side; as viewed from this angle, the crater itself looks like a deep blue crevice at bottom. North is up.

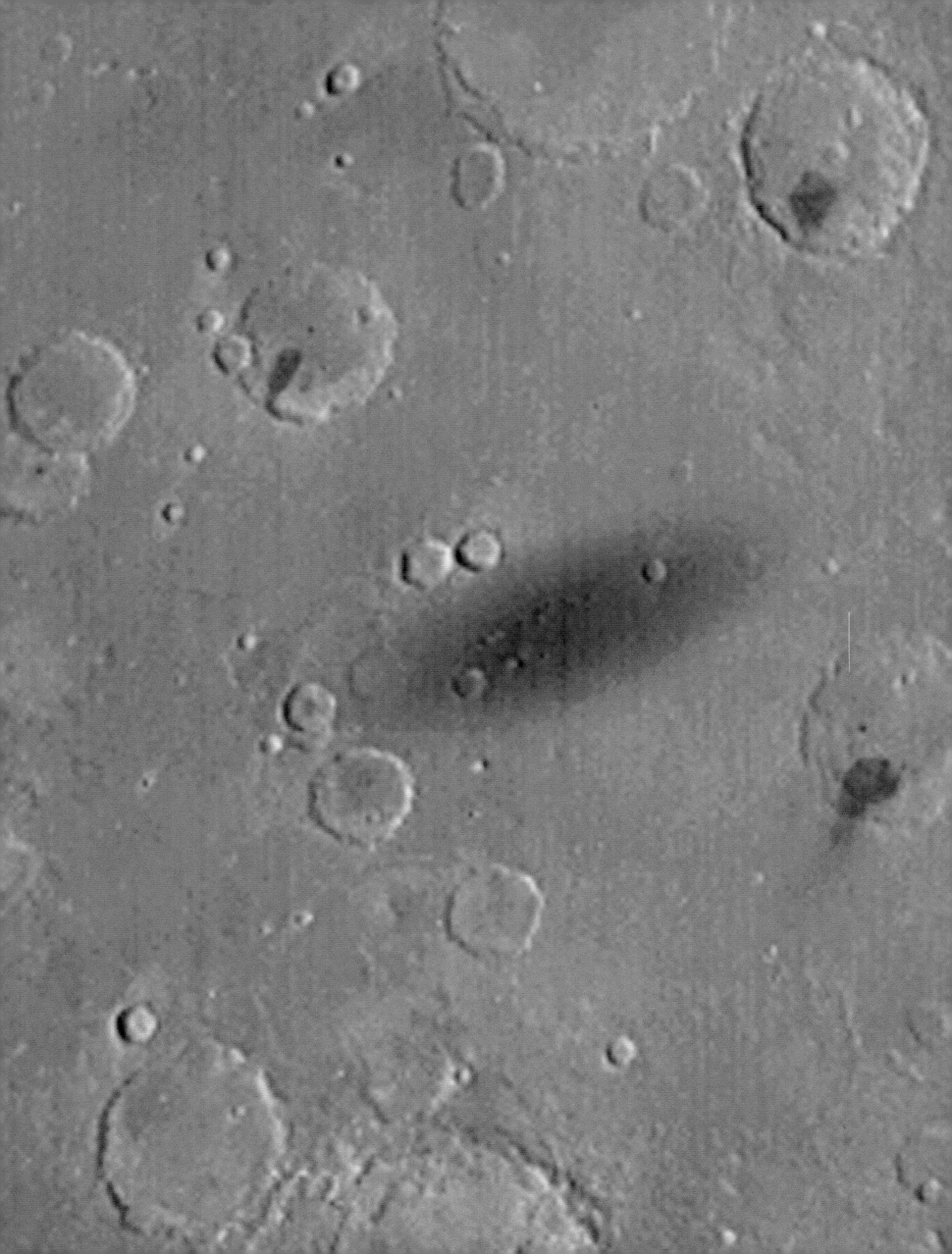

instead space stations abandoned by a deceased Martian civilization, something even the imaginative Percival Lowell had never dreamed up. Wrote Shklovskii, "The idea that the moons of Mars are artificial satellites may seem fantastic, at first glance. In my opinion, however, it merits serious consideration." Shklovskii also believed that advanced extraterrestrial civilizations were communist, since capitalist planets would blow themselves up.

In the case of the Martian moons, Shklovskii reasoned, if that is the right word, as follows. He said that the orbit of Phobos was decaying because the moon rubbed against the Martian atmosphere. For atmospheric friction to produce the observed decay rate, he calculated that Phobos had to be no more than 1/1000 as dense as water, suggesting that the moon was hollow. This, in turn, implied that Phobos and its comrade, Deimos, were artificial objects. But when American spacecraft first imaged the two moons, one official joked that if they were space stations, the Martians had disguised them well.

Phobos hasn't long to live. Modern observations indicate that its distance from the red planet shrinks 1 to 2 inches a year—not because of atmospheric friction but because Mars and Phobos exert tidal forces on each other. Unless human beings intervene to save the moon, it will smash into the Martian surface within 40 million years. How many other Martian moons have already met the same fate? Deimos is safe, however, because it's farther away. Indeed, theory suggests that Deimos is slowly receding from Mars, although this has not yet been observed.

To a colonist on the Martian surface, neither satellite would impress as much as the Moon does

FACING PAGE: Phobos casts an elliptical shadow onto Mars.

Both moons offer a wonderful view of the world below

from the Earth. That's because the moons, though close to Mars, are so small. Phobos would appear one third as large as the Moon and about one twentieth as bright. Deimos would look like a star, but far brighter than Venus. Because of the moons' proximity— Phobos is just 1/63 as far from the Martian surface as the Moon is from Earth's surface—their dark sides would glow. This glow comes from sunlight bouncing off Mars. Likewise, reflected light from the Earth, called earthshine, illuminates the lunar dark side, but marsshine is far more intense.

Phobos and Deimos move opposite ways through the Martian sky. Phobos whizzes from west to east, rising and setting two or three times a day, whereas Deimos lingers, requiring nearly three days to progress from moonrise to moonset. Since both moons huddle close to Mars and orbit close to its equatorial plane, no colonist near the poles of Mars would see either moon at all.

In the future, the moons may serve as space stations to support Martian colonists, since their low escape velocities allow an easy launch to the Martian surface. Furthermore, both moons offer a wonderful view of the world below: a desert planet with towering volcanoes, giant impact basins, enormous canyons, globe-encircling dust storms, and the future abode of adventurous terrestrial life.

PLANET DATA

	EARTH		
	Mean Distance	Minimum Distance	Maximum Distance
Distance from Sun's Center*	1.0000 AU 92,956,000 miles 149,600,000 km 8.3167 light-min	0.9833 AU 91,402,000 miles 147,100,000 km 8.1778 light-min	1.0167 AU 94,509,000 miles 152,100,000 km 8.4557 light-min
Sidereal Year	365.256 days**		
Day—Sidereal —Synodic	23 hours 56 minutes 4 seconds 24 hours		
Axial Tilt	23.44 degrees		
Equatorial Diameter	7,926.4 miles 12,756.3 kilometers		
Orbital Eccentricity	1.67 percent		
Orbital Inclination to Earth's Orbit	0 degrees		
Mean Orbital Speed Around Sun	66,629 miles per hour 29.79 kilometers per second		
Mass (Earth = 1)	1.000		
Mean Density (water = 1)	5.52 grams per cubic centimeter		
Equatorial Surface Gravity (Earth = 1)	1.00		
Equatorial Escape Velocity	25,010 miles per hour 11.18 kilometers per second		
Mean Sunlight Intensity (Earth = 1)	1.00		
Mean Surface Temperature	+60 Fahrenheit +16 Celsius 289 Kelvin		
Mean Atmospheric Pressure	29.92 inches of mercury 1,013.25 millibars		
Atmospheric Composition	78% nitrogen, 21% oxygen		
Number of Moons	1		
Symbol	⊕		
Noteworthy Accomplishments	Life Oceans Big moon		

* One AU—astronomical unit—is the mean Sun-Earth distance, 92,955,807 miles or 149,597,871 kilometers.
   One light-minute is the distance light travels in one minute.

** One day is 24 hours.

MARS

Mean Distance	Minimum Distance	Maximum Distance
1.5237 AU	1.3813 AU	1.6660 AU
141,630,000 miles	128,400,000 miles	154,860,000 miles
227,940,000 km	206,640,000 km	249,230,000 km
12.672 light-min	11.488 light-min	13.856 light-min

686.98 days**

24 hours 37 minutes 23 seconds
24 hours 39 minutes 35 seconds

25.19 degrees

4,220.6 miles
6,792.4 kilometers

9.34 percent

1.85 degrees

53,979 miles per hour
24.13 kilometers per second

0.1074

3.94 grams per cubic centimeter

0.38

11,200 miles per hour
5.02 kilometers per second

0.43

−67 Fahrenheit
−55 Celsius
218 Kelvin

0.2 inch of mercury
6 millibars

95% carbon dioxide, 3% nitrogen

2

♂

Solar system's tallest mountains
Solar system's largest canyons

MOON DATA

	MOON	PHOBOS	DEIMOS
Satellite of	Earth	Mars	Mars
Discoverer	———————	Asaph Hall	Asaph Hall
Discovery Date*	———————	August 18, 1877	August 12, 1877
Mean Distance from Planet's Center	238,860 miles 384,400 kilometers	5,827 miles 9,378 kilometers	14,577 miles 23,459 kilometers
Mean Distance from Planet's Surface	234,890 miles 378,020 kilometers	3,717 miles 5,982 kilometers	12,466 miles 20,063 kilometers
Orbital Period	27 days 7 hours 43 minutes 12 seconds	7 hours 39 minutes 14 seconds	30 hours 17 minutes 55 seconds
Rotation Period	same	same	same
Diameter	2,159 miles 3,475 kilometers	16.1 x 13.7 x 11.5 miles (Mean: 13.7 miles) 25.9 x 22.0 x 18.4 kilometers (Mean: 22.1 kilometers)	9.7 x 7.5 x 6.3 miles (Mean: 7.8 miles) 15.6 x 12.0 x 10.2 kilometers (Mean: 12.6 kilometers)
Orbital Eccentricity	5.49 percent	1.5 percent	0.02 percent
Orbital Inclination to Planet's Equator	18.28 to 28.58 degrees	1.1 degrees	0.9 to 2.7 degrees
Mass (Earth = 1)	0.0123000	0.000000002 (2×10^{-9})	0.0000000004 (4×10^{-10})
Mean Density (water = 1)	3.35 grams per cubic centimeter	1.9 grams per cubic centimeter	1.8 grams per cubic centimeter
Albedo (Amount of Sunlight Reflected)	12 percent	7 percent	7 percent

* Eastern Time.

ATMOSPHERES

	VENUS	EARTH	MARS
	(abundance by volume in percent)		
Carbon Dioxide (CO_2)	96.5	0.037 and rising	95.32
Nitrogen (N_2)	3.5	78.08	2.7
Argon (Ar)	0.007	0.934	1.6
Oxygen (O_2)	0.001	20.95	0.13
Carbon Monoxide (CO)	0.0045	0.000015	0.07
Water Vapor (H_2O)	0.0045	up to 4	0.03
Hydrogen (H_2)	————	0.00005	0.0015
Helium (He)	0.0012	0.000524	0.0004
Neon (Ne)	0.0007	0.00182	0.00025
Krypton (Kr)	0.00002	0.000114	0.00003
Ozone (O_3)	————	0.000004	0.00001
Xenon (Xe)	————	0.0000087	0.000008
Methane (CH_4)	————	0.000172	————
Sulfur Dioxide (SO_2)	0.01	0.00000003	————
Mean Surface Pressure (bars)	93	1.01325	0.006

THE LIFE OF MARS

PERIOD	YEARS AGO (BILLIONS)	EARTH	AIR
Early Noachian	4.6 to 3.9	Heavy bombardment from asteroids and comets Iron core forms within 13 million years of planet's birth Magnetic field strengthens Oldest Martian meteorite, ALH 84001 (4.5 billion years old) Continental drift? Iron core cools; magnetic field dies Impacts create Hellas and Argyre basins	Small impacts deliver gas; large impacts remove it Impacting comets endow Mars with a massive hydrogen atmosphere? Hydrogen escapes into space? Drags heavier gases with it, removing most of original atmosphere? Magnetic field protects remaining atmosphere from solar wind Greenhouse effect warms Mars to above freezing? Life emits methane and enhances greenhouse effect? Solar wind begins to strip atmosphere of gas
Middle Noachian	3.9 to 3.8	Heavy bombardment from asteroids and comets continues	Impacts and solar wind continue to strip atmosphere
Late Noachian	3.8 to 3.6	Heavy bombardment from asteroids and comets ends	Impact erosion of atmosphere ends; solar-wind stripping continues
Early Hesperian	3.6 to 3.5	Northern plains are flooded with lava	Climate deteriorates as air thins and greenhouse effect dwindles
Late Hesperian	3.5 to 3.1*	Floods of water carry sediment into northern plains, burying the lava there	
Early Amazonian	3.1* to 1.8*		
Middle Amazonian	1.8* to 0.4	Most Martian meteorites	Climate cycles alternately thin and thicken air; bringing brief periods of warmth?
Late Amazonian	0.4 to 0	Youngest Martian meteorites	Climate cycles continue Frequent dust storms

* This number is especially uncertain.

FIRE	WATER
Impacts and the formation of the planet's iron core melt the planet; the planet's surface likely sports a magma ocean	Bombarding asteroids and comets deliver water to Mars
Volcanoes erupt throughout the southern highlands	
Volcanoes vent gases to the atmosphere, especially carbon dioxide and water vapor	Rivers of water flow over the surface Rainfall? Lakes in craters? An ocean occupies the northern plains? Water escapes into space
Tharsis bulge forms; planet pirouettes in response?	Lakes likely in Hellas and Argyre impact basins Rivers continue to flow Northern ocean?
Volcanoes in Tharsis and elsewhere continue to emit greenhouse gases Highland volcanism decreases	Northern ocean? Rivers continue to flow Water continues to escape into space
Valles Marineris canyons form Tharsis and Elysium volcanic eruptions flood northern plains with lava	Rivers begin to dry up; some remain near volcanoes Water continues to escape into space
	Floods spill into northern plains; creating temporary ocean? Most remaining water freezes at poles or underground?
Volcanic eruptions decline	Occasional floods
Occasional eruptions from volcanoes in Tharsis and Elysium; melting enough ice to create a temporary ocean?	
Occasional volcanic eruptions	Water produces gullies

TABLE IV

NASA MISSIONS TO THE PLANETS

NAME	LAUNCH DATE*	TARGET	TYPE
Mariner 1	July 22, 1962	Venus	Flyby
Mariner 2	August 27, 1962	Venus	Flyby
Mariner 3	November 5, 1964	Mars	Flyby
Mariner 4	November 28, 1964	Mars	Flyby
Mariner 5	June 14, 1967	Venus	Flyby
Mariner 6	February 24, 1969	Mars	Flyby
Mariner 7	March 27, 1969	Mars	Flyby
Mariner 8	May 8, 1971	Mars	Orbiter
Mariner 9	May 30, 1971	Mars	Orbiter
Pioneer 10	March 2, 1972	Jupiter	Flyby
Pioneer 11	April 5, 1973	Jupiter	Flyby
		Saturn	Flyby
Mariner 10	November 3, 1973	Venus	Flyby
		Mercury	Flyby
		Mercury	Flyby
		Mercury	Flyby
Viking 1	August 20, 1975	Mars	Orbiter
		Mars	Lander
Viking 2	September 9, 1975	Mars	Orbiter
		Mars	Lander
Pioneer Venus Orbiter	May 20, 1978	Venus	Orbiter
Pioneer Venus Multiprobe	August 8, 1978	Venus	Landers
Voyager 1	September 5, 1977	Jupiter	Flyby
		Saturn	Flyby
Voyager 2	August 20, 1977	Jupiter	Flyby
		Saturn	Flyby
		Uranus	Flyby
		Neptune	Flyby
Magellan	May 4, 1989	Venus	Orbiter
Mars Observer	September 25, 1992	Mars	Orbiter
Galileo	October 18, 1989	Jupiter	Orbiter
		Jupiter	Probe
Mars Pathfinder	December 4, 1996	Mars	Lander
Mars Global Surveyor	November 7, 1996	Mars	Orbiter
Mars Climate Orbiter	December 11, 1998	Mars	Orbiter
Mars Polar Lander	January 3, 1999	Mars	Lander
Mars Odyssey	April 7, 2001	Mars	Orbiter
Mars Spirit	June 10, 2003	Mars	Lander
Mars Opportunity	July 7, 2003	Mars	Lander
Cassini	October 15, 1997	Saturn	Orbiter

* Eastern Time.

** Time at spacecraft in Eastern Time—transmission to Earth took minutes to hours, depending on target's distance from Earth.

SUCCESS?	ENCOUNTER DATE**	NOTES
No	————	Veered off course after launch; blown up
Yes	December 14, 1962	First successful planetary mission; first Venus flyby
No	————	Shroud failure
Yes	July 14, 1965	First successful Mars mission
Yes	October 19, 1967	
Yes	July 31, 1969	
Yes	August 5, 1969	
No	————	Launch failure
Yes	November 13, 1971	First successful Mars orbiter
Yes	December 3, 1973	First traversal of asteroid belt; first Jupiter flyby
Yes	December 3, 1974	
Yes	September 1, 1979	First Saturn flyby
Yes	February 5, 1974	
Yes	March 29, 1974	First Mercury flyby
Yes	September 21, 1974	
Yes	March 16, 1975	
Yes	June 19, 1976	
Yes	July 20, 1976	First successful Mars lander
Yes	August 7, 1976	
Yes	September 3, 1976	
Yes	December 4, 1978	
Yes	December 9, 1978	
Yes	March 5, 1979	
Yes	November 12, 1980	
Yes	July 9, 1979	
Yes	August 25, 1981	
Yes	January 24, 1986	First Uranus flyby
Yes	August 24, 1989	First Neptune flyby
Yes	August 10, 1990	
No	————	Failure on approach to Mars
Yes	December 7, 1995	First Jupiter orbiter
Yes	December 7, 1995	First probe to enter Jupiter's atmosphere
Yes	July 4, 1997	
Yes	September 11, 1997	
No	————	Passed too close to Mars; burned up in atmosphere
No	————	Crashed on Mars
Yes	October 23, 2001	

Glossary

Amazonian The most recent period of Martian history, encompassing roughly two thirds of the planet's life.

Andesite A silicon-rich volcanic rock often associated with continental drift. Its presence on the Martian surface, however, may merely indicate that other processes melted parts of the surface; this melting boosted the silicon content.

Axial Tilt The angle by which a planet's rotation axis deviates from perfect uprightness—that is, the angle between the rotation axis and the vertical to the planet's orbit. The Earth and Mars have similar axial tilts, 23.4 and 25.2 degrees, respectively. The axial tilts cause the planets' seasons.

Basalt A dark volcanic rock high in magnesium and iron, and the most common igneous rock on the Earth, the Moon, and probably Mars.

Carbonate A mineral such as calcite or a rock such as limestone based on carbon and oxygen. Carbonates precipitate from a lake or ocean; their apparent absence from Mars poses problems for the idea that the planet once had large bodies of water.

Core The central part of a planet. The cores of Mercury, Venus, Earth, and Mars are mostly iron; above the core are the mantle and crust.

Crust The topmost layer of a planet, above the core and mantle.

Eccentricity A measure of how round or oval an orbit is. A perfect circle has an eccentricity of 0 percent, but the more elliptical the orbit, the greater is its eccentricity. The eccentricity of the Earth's orbit is 1.7 percent; that of the Martian orbit is over five times greater, 9.3 percent, so Mars has a much more elliptical orbit than the Earth.

Greenhouse Effect The trapping of warmth by gases in a planet's atmosphere: visible sunlight passes through the gases and heats the planet's surface, which radiates the heat at infrared wavelengths that the gases block. Greenhouse gases include carbon dioxide, water vapor, and ozone. On Mars the greenhouse effect raises the temperature by 10 degrees F., on Earth by 60 degrees F., and on Venus by 900 degrees F.

Heavy Bombardment The period during the first 800 million years of the solar system's life when the planets suffered a large number of impacts from asteroids and comets.

Hematite An iron oxide that can form in hot water. Its presence on Mars, in Meridiani Planum and elsewhere, suggests that these regions once had hot water that possibly sparked life.

Hesperian The middle period of Martian history. Younger is the Amazonian; older is the Noachian.

Inclination The angle that a planet's orbit makes with the Earth's orbit. Mars's orbital inclination is only 1.85 degrees.

Mantle The large region surrounding a planet's core. The mantles of Earth and Mars constitute most of each planet and consist mainly of silicates.

Noachian The oldest period of Martian history: the first billion years of the planet's life, including the heavy bombardment and its immediate aftermath.

Opposition The ideal time to observe Mars, when the planet is on the opposite side of the Earth from the Sun. As a result, it's visible all night long: the planet rises around sunset and sets around sunrise. Martian oppositions occur approximately every two years and two months; especially favorable oppositions occur every fifteen or seventeen years.

Outgassing The venting of gases from a planet's interior to the atmosphere, often via volcanic eruptions. Because Mars has little volcanic activity, its present outgassing rate is only a fraction of Earth's.

Planetesimal A rocky, asteroid-like body that collided with others to build the planets.

Precession The wobbling of a planet that causes its axis to point in different directions at different times. Precession causes the Earth's axis to wobble with a period of 23,000 years and the Martian axis with a period of 51,000 years.

Revolution The motion of a planet around the Sun. The Earth has a revolution period of 1 year and Mars 1.88 years.

Rotation The spinning of a planet. The Earth and Mars have similar rotation periods, around 24 hours.

Sapping The seepage of groundwater, which erodes valleys. Sapping rather than rainfall may have created the riverbeds on Mars.

Sidereal Rotation Period The true rotation period of a planet, which an external observer would see. The Earth's sidereal rotation period is 23 hours, 56 minutes; that of Mars is 24 hours, 37 minutes. Compare **Synodic Rotation Period.**

Silicate A rock or mineral based on silicon and oxygen. The crust and mantle of the Earth and Mars are largely silicate.

Solar Wind A stream of charged particles from the Sun that tears away the atmospheres of planets, such as Mars, that have no magnetic field to protect themselves.

Stratigraphy The study of geological strata and on Mars the assigning of different time periods—Amazonian, Hesperian, or Noachian—to various regions of the planet.

Synodic Rotation Period The rotation period of a planet as seen by someone living on the planet—in particular, the average time from one noon to the next. This differs from the true rotation period because as the planet spins it is also going around the Sun. The Earth's synodic rotation period is 24 hours; that of Mars is 24 hours, 40 minutes. Compare **Sidereal Rotation Period.**

Further Reading

Barbree, Jay, Caidin, Martin, and Wright, Susan, 1997. *Destination Mars* (New York: Penguin).

Bergreen, Laurence, 2000. *Voyage to Mars* (New York: Riverhead).

Blunck, Jürgen, 1982. *Mars and Its Satellites*, second edition (Smithtown, New York: Exposition Press).

Boyce, Joseph, 2002. *The Smithsonian Book of Mars* (Washington, D.C.: Smithsonian Institution Press).

Carr, Michael H., 1981. *The Surface of Mars* (New Haven, Connecticut: Yale University Press).

———, 1996. *Water on Mars* (Oxford: Oxford University Press).

Cattermole, Peter John, 2001. *Mars: The Mystery Unfolds* (Oxford: Oxford University Press).

Godwin, Robert (editor), 2000. *Mars: The NASA Mission Reports* (Burlington, Ontario: Apogee).

Goldsmith, Donald, 1997. *The Hunt for Life on Mars* (New York: Dutton).

Hartmann, William K., 2003. *A Traveler's Guide to Mars* (New York: Workman).

Hoyt, William Graves, 1976. *Lowell and Mars* (Tucson: University of Arizona Press).

Kallenbach, Reinald, Geiss, Johannes, and Hartmann, William K. (editors), 2001. *Chronology and Evolution of Mars* (Dordrecht: Kluwer).

Kieffer, Hugh H., Jakosky, Bruce M., Snyder, Conway W., and Matthews, Mildred S. (editors), 1992. *Mars* (Tucson: University of Arizona Press).

Lowell, Percival, 1895. *Mars* (Boston: Houghton Mifflin).

———, 1906. *Mars and Its Canals* (New York: Macmillan).

———, 1908. *Mars as the Abode of Life* (New York: Macmillan).

Moore, Patrick, 1998. *Patrick Moore on Mars* (London: Cassell).

Morton, Oliver, 2002. *Mapping Mars* (New York: Picador).

Raeburn, Paul, 1998. *Mars: Uncovering the Secrets of the Red Planet* (Washington, D.C.: National Geographic Society).

Sheehan, William, 1996. *The Planet Mars* (Tucson: University of Arizona Press).

Sheehan, William, and O'Meara, Stephen James, 2001. *Mars: The Lure of the Red Planet* (Amherst, New York: Prometheus).

Shirley, Donna, and Morton, Danelle, 1998. *Managing Martians* (New York: Broadway).

Strauss, David, 2001. *Percival Lowell* (Cambridge, Massachusetts: Harvard University Press).

VanDecar, John (editor), 2001. "Nature Insight: Mars," in *Nature*, 412, 207 (July 12, 2001).

Walter, Malcolm, 1999. *The Search for Life on Mars* (Cambridge, Massachusetts: Perseus).

Wilford, John Noble, 1990. *Mars Beckons* (New York: Knopf).

Zubrin, Robert, and Wagner, Richard, 1996. *The Case for Mars* (New York: The Free Press).

Illustration Credits

JPL stands for the Jet Propulsion Laboratory, and NASA, of course, for the National Aeronautics and Space Administration. All images in this book were digitally reprocessed by Tony Hallas.

PAGE	OBJECT	CREDIT
Mars Through the Centuries		
4	Mars Upside Down	Hubble Space Telescope. Steve Lee (University of Colorado), Jim Bell (Cornell University), Mike Wolff (Space Science Institute), and NASA.
8	Earth	Apollo 17. NASA.
9	Mars to Scale	Viking Orbiter. NASA/JPL/U.S. Geological Survey.
15	Percival Lowell	Copyright © Lowell Observatory Archives.
16	Lowell Crater	Mars Global Surveyor. NASA/JPL/Malin Space Science Systems.
21	Odyssey Launch	Kennedy Space Center/NASA.
24	Crescent Mars	Viking 2 Orbiter. NASA/JPL.
Earth		
26	Yogi	Mars Pathfinder. NASA/JPL.
29	Amazonis Planitia	Viking Orbiter. NASA/JPL/U.S. Geological Survey.
31	Hesperia Planum	Viking Orbiter. NASA/JPL/U.S. Geological Survey.
32	Noachis Terra	Viking Orbiter. NASA/JPL/U.S. Geological Survey.
35	Mars, Longitude 20°W	Hubble Space Telescope. Steve Lee (University of Colorado), Todd Clancy (Space Science Institute, Boulder, Colorado), Phil James (University of Toledo), and NASA.
37	Mars, Longitude 94°W	Hubble Space Telescope. Phil James (University of Toledo), Todd Clancy (Space Science Institute, Boulder, Colorado), Steve Lee (University of Colorado), and NASA.
39	Mars, Longitude 160°W	Hubble Space Telescope. David Crisp and the WFPC2 Science Team (JPL/California Institute of Technology), and NASA.
41	Mars, Longitude 210°W	Hubble Space Telescope. David Crisp and the WFPC2 Science Team (JPL/California Institute of Technology), and NASA.
43	Mars, Longitude 288°W	Hubble Space Telescope. Steve Lee (University of Colorado), Todd Clancy (Space Science Institute, Boulder, Colorado), Phil James (University of Toledo), and NASA.
45	Mars, Longitude 305°W	Hubble Space Telescope. David Crisp and the WFPC2 Science Team (JPL/California Institute of Technology), and NASA.
48	Full Mars	Mars Global Surveyor. Image: Michael Caplinger and Michael Malin, Malin Space Science Systems; data: NASA/JPL/Malin Space Science Systems/Mars Orbiter Laser Altimeter (MOLA) Science Team.
50	Mars Full Topography	Mars Global Surveyor. NASA/JPL/MOLA Science Team.
54	Mars North Polar Topography	Mars Global Surveyor. NASA/JPL. Map prepared by the U.S. Geological Survey for NASA, using data from the MOLA Science Team.

PAGE	OBJECT	CREDIT
Water		
120	Scamander Vallis	Viking Orbiter. NASA/JPL/U.S. Geological Survey.
125	Ares Vallis	Mars Odyssey. NASA/JPL/Arizona State University.
126	Terra Sabaea Riverbeds	Viking Orbiters. NASA/JPL/U.S. Geological Survey.
129	Nanedi Vallis	Mars Global Surveyor. NASA/JPL/Malin Space Science Systems.
130	Newton Crater Gullies	Mars Global Surveyor. NASA/JPL/Malin Space Science Systems.
131	Noachis Terra Gullies	Mars Global Surveyor. NASA/JPL/Malin Space Science Systems.
134	Hellas	Viking Orbiter. NASA/JPL/U.S. Geological Survey.
136	Hellas Topographic Hemisphere	Mars Global Surveyor. NASA/JPL/MOLA Science Team.
137	Hellas Valleys	Mars Global Surveyor. NASA/JPL/Malin Space Science Systems.
138	Argyre	Mars Global Surveyor. NASA/JPL/MOLA Science Team and G. Shirah, NASA Goddard Space Flight Center Scientific Visualization Studio.
139	Sedimentary Layers	Mars Global Surveyor. NASA/JPL/Malin Space Science Systems.
140	Ma'adim Vallis/Gusev Crater	Viking Orbiter. NASA/JPL/U.S. Geological Survey.
143	Topographic Ocean I	Mars Global Surveyor. NASA/JPL/MOLA Science Team.
144	Topographic Ocean II	Mars Global Surveyor. NASA/JPL/MOLA Science Team.
145	Topographic Ocean III	Mars Global Surveyor. NASA/JPL/MOLA Science Team.
146	Topographic Ocean IV	Mars Global Surveyor. NASA/JPL/MOLA Science Team.
149	Frost	Viking 2 Lander. NASA/JPL.
151	North Polar Cap	Mars Global Surveyor. NASA/JPL/Malin Space Science Systems.
152	South Polar Cap	Mars Global Surveyor. NASA/JPL/Malin Space Science Systems.
154	Odyssey Ice Map	Mars Odyssey. NASA/JPL/University of Arizona/Los Alamos National Laboratories.
156	East Mangala Vallis Splosh Crater	Viking Orbiter. NASA/JPL/U.S. Geological Survey.
159	Comet Hale-Bopp	Copyright © Kenneth Lum.
162	Quintuplet Cluster	Hubble Space Telescope. Don Figer (Space Telescope Science Institute) and NASA.
164	Face on Mars, small	Viking Orbiter. NASA/JPL.
165	Face on Mars, large	Mars Global Surveyor. NASA/JPL/Malin Space Science Systems.
166	Galle Crater	Mars Global Surveyor. NASA/JPL/Malin Space Science Systems.
The Moons of Mars		
170	Phobos with Mars	Soviet Phobos mission. G. Avenasov and B. Zhukov, IKI, USSR; color processing by Mark Robinson (Northwestern University) and Fraser Fanale (University of Hawaii) in collaboration with the Soviet Phobos imaging team.
173	Phobos	Viking Orbiter. NASA/JPL. Damon Simonelli (JPL) and Joseph Veverka (Cornell University).
175	Phobos Close-up	Mars Global Surveyor. NASA/JPL/Malin Space Science Systems.
176	Deimos	Viking Orbiter. NASA/JPL. Courtesy Peter Thomas (Cornell University).
179	Phobos Topography	From Viking Orbiter data. Peter Thomas (Cornell University).
180	Deimos Topography	From Viking Orbiter data. Peter Thomas (Cornell University).
182	Phobos Shadow	Mars Global Surveyor. NASA/JPL/Malin Space Science Systems.

Index

Mars *(cont.)*

ABOUT THE AUTHOR

Ken Croswell is the author of several highly acclaimed books, including *Magnificent Universe* and *See the Stars: Your First Guide to the Night Sky*. He earned his Ph.D. in astronomy from Harvard University and lives in Berkeley, California.